Evolutionary Psychology as Maladapted Psychology

Life and Mind: Philosophical Issues in Biology and Psychology
Kim Sterelny and Robert A. Wilson, editors

Evolutionary Psychology as Maladapted Psychology

Robert C. Richardson

A Bradford Book
The MIT Press
Cambridge, Massachusetts
London, England

MIT Press books may be purchased at special quantity discounts for business or sales promotional use. For information, please e-mail special_sales@mitpress.mit.edu or write to Special Sales Department, The MIT Press, 55 Hayward Street, Cambridge, MA 02142.

This book was set in Times Roman and Syntax by SNP Best-set Typesetter Ltd., Hong Kong, and was printed and bound in the United States of America.

Library of Congress Cataloging-in-Publication Data

Richardson, Robert C., 1949–.
Evolutionary psychology as maladapted psychology / by Robert C. Richardson.
    p.   cm.—(Life and mind)
"A Bradford book."
Includes bibliographical references and index.
ISBN 978-0-262-18260-7 (hardcover : alk. paper)
1. Evolutionary psychology. I. Title.

BF698.95.R44   2007
155.7—dc22   2006030807

10 9 8 7 6 5 4 3 2 1

Contents

Preface and Acknowledgments

Evolutionary psychology is by nature a hybrid discipline. The very name requires that it at least pay attention to evolutionary biology as one mistress, and to psychology as another. Philosophy might seem the odd one out. Locke thought of philosophy as a handmaiden to science. I think of philosophy as a facilitator. In this case, there is a discussion to be promoted.

I am by inclination and by profession a philosopher of science, interested in doing philosophy of science from the inside, engaging the details of the science, rather than from the outside, pretending to impose some independent standard on the sciences. Thus practiced, philosophy of science is a hybrid discipline. I work both within philosophy of biology and within philosophy and cognitive science. An outsider's philosophical perspective—which promotes a normative standard apart from the practice of science—was once common in philosophy of science. It is still lamentably common within much of philosophy; it isn't common any longer within philosophy of science. Philosophy of biology has generally abandoned any pretense of a stance that issues some sort of "standard" apart from the practice of biologists. Philosophy of mind, however, has a more ambiguous status. Some still maintain an independent stance. They think there is some standard of evidence apart from, and prior to, psychological practice. I do not traffic in the *a priori*. My interests in cognitive science, as in philosophy of biology, respect the science. This particular book pays more heed to the norms within biology than within psychology. David Buller, by contrast, is a philosopher who has offered a methodological critique of evolutionary psychology from a psychological perspective. I'm largely sympathetic with that critique. I don't defend its details, but I think the invitation to a more reflective methodology is salutary for evolutionary psychology. Even if evolutionary psychologists resist his conclusions, they should at least answer the problems of method he raises. Philip Kitcher is another philosopher who has raised a series of issues concerning sociobiology and, more recently, evolutionary psychology. Once again, I'm largely sympathetic

Preface and Acknowledgments

with his critique of sociobiology. And, again, I think the methodological issues he raises need addressing, whether or not one accepts his substantive critique.

My concerns are very much in harmony with both Buller and Kitcher. Buller focuses on the *psychological* credentials of evolutionary psychology. Like Kitcher, I focus on the *evolutionary* credentials of evolutionary psychology. As a philosopher, my concern is primarily with issues methodological. I am interested, first and foremost, in what we would need to know in order to validate the claims of evolutionary psychology. In particular, I am interested in what we would need to know in order to vindicate the *evolutionary* claims of evolutionary psychology. I sometimes describe these as *evolutionary pretensions*, because they are not explicitly argued for so much as assumed. They are part of the rhetoric. I take the pretensions seriously, exploring the various avenues available for empirically validating specific evolutionary claims, and asking how well the literature in evolutionary psychology fares against those standards.

My first ventures into the topics raised here were in the context of a seminar within the Department of Biological Sciences at the University of Cincinnati in the late 1970s. It was a robust discussion of E. O. Wilson's *Sociobiology*, and of the various models which lay behind it, looking at it chapter by chapter. It was an engaging experience. I gained a grounding in what we might learn about social behavior from an evolutionary perspective. I still think there is a great deal to be learned, and a great deal that has been learned. Most of our discussion then concerned ants, spiders, and occasionally hyenas and lions; our discussion was only incidentally about human behavior, as for that matter was Wilson's book. In the same period, I was discussing similar issues with friends and colleagues in the Department of Psychology at the University of Cincinnati. Some of this discussion concerned the implications of evolutionary biology for psychology. More often, the topics were more focused on psychology. Of course, here the discussion was less with spiders than with such things as incest avoidance. My good fortune continues still with colleagues in both departments. My life is enriched by all of them. For those who live out their academic lives within the confines of one department, it's difficult to imagine how rewarding this kind of interaction can be. Without wanting for a minute to diminish the appreciation gained from knowing something in exquisite detail, I have also gained much from the interaction with my peers. So to my various colleagues in the Department of Biological Sciences and the Department of Psychology at the University of Cincinnati, I am especially grateful.

My interests in the topics at the intersection of evolutionary biology and philosophy did not wane over the ensuing decades, although, as I've said, I did not formally enter the discussion. I certainly did not plan on writing an extended piece on evolutionary psychology. It seems to have been more something that happened to me. With my twin interests in evolutionary biology and cognitive psychology, perhaps it was inevitable that I at least engage the discussion. At the invitation of James Fetzer and Paul Davies, I did enter the discussion about ten years ago. At that point, I thought of the topic as a diversion from my main interests. It was an interesting diversion, but a diversion nonetheless. In the years that followed, I maintained an interest in the topic, and what started as a diversion assumed a kind of structure, and a life of its own. I ended up writing a series of papers on evolutionary psychology, all engaged with asking how the evolutionary pretensions of evolutionary psychology could be grounded. That is, my question was how we could know what evolutionary psychologists claim to know in order to get their psychology off the ground. I discussed my worries with colleagues in psychology and in biology. Much to my surprise, my skepticism was shared by colleagues both in biology and psychology, as it was by my colleagues in philosophy. I am sure my skepticism will be less warmly greeted by advocates of evolutionary psychology.

Following my initial foray into issues concerned more directly with evolutionary psychology at the invitation of Fetzer and Davies, I found there was an interest among others in approaching these issues from the point of view of philosophy of science. As a consequence, I was invited to give talks on evolutionary psychology at a number of universities and societies over this period. Since I did not want to do the same thing over and over, I began to branch out, though working within the same theme. Soon it seemed there was a kind of system to the madness. (That would be my madness, not the madness of evolutionary psychology.) By the time there was a series of publications, several of my friends were urging me to do a book. When there's a system to the madness, that constitutes a book. When there's not, I guess it's a collection of articles. This book is not a collection of articles. It draws on, and elaborates on, the articles I have written on the topics over the last decade. This volume develops the themes of the various articles and presentations, but without reprinting them. Thankfully, my views have not remained static over the period. Had it not been for the opportunity provided by writing these articles, and the opportunity to talk about the issues, I surely would not have produced this book. I am grateful for the various audiences, and for their input.

One of the serendipitous results of this project is that it has encouraged me, as a philosopher of science, to think more systematically about the place of natural selection and its alternatives in evolutionary theory. It has also forced me to think more about how we distinguish the alternatives empirically. That forms the backbone of the book, which is structured around three ways we can approach questions concerning the role of natural selection within evolutionary theory. For philosophers of biology, that theme, along with the case studies I offer, might be of more enduring interest.

During the period I've been working on the book, I've also gained from discussions with colleagues in cognitive science in the Netherlands and in Germany at the University of Osnabrück. I was fortunate enough, during this period, to have visiting appointments at the Free University of Amsterdam within the Department of Molecular Cell Physiology, and at the University of Osnabrück within the Department of Cognitive Science. They were, inevitably, subjected to my interest in evolutionary psychology even though our joint interests were far removed from that topic.

Three people have read or commented on all of the manuscript. Each changed the book in substantial ways. One is Paul Davies, who was a postdoc at the University of Cincinnati many years ago. I count him as one of my best friends, and his comments changed the manuscript substantially. Another is Stephen Downes, whom I also count as a friend, and who focused critically on the thinking about adaptation. Finally, Michael Bailey offered some incisive critical commentary, particularly on the interpretation of the key idea of heritability. I think that I have incorporated many of their insights. I have certainly benefited from them, though of course that does not at all imply that they agree with much of what is included here. Each has, in any case, improved the book.

General debts are one thing. It would be remiss to avoid specific debts. Some are acknowledged in the text that follows. I am sure that many have made contributions which I have incorporated while unintentionally suppressing the contributor. I am also sure that many have offered contributions that I have failed to incorporate, sometimes because I've disagreed and sometimes because I've not properly appreciated the point. Still, following the artificial divisions that define academic disciplines, I need at least to acknowledge the following individuals. Within biology, Maricia Bernstein, Fred Boogert, Frank Bruggeman, Rebecca German, Richard Lewontin, George Uetz, and Wim van der Steen have been significant in shaping my intellectual agenda. In psychology and cognitive science, I've profited especially from discussions with George Bishop, William Dember, Huib Looren de Jong, Maurice Schouten, Don Schumsky, and Dan Wheeler. Within philosophy, I've gained from various discussions with William Bechtel, John Bickle, Robert Brandon, Richard Burian, Christine Cuomo, Marjorie Greene, Donald Gustafson, Lynne Hankinson, Lawrence Jost, John McEvoy, W. E. Morris, Thomas Polger, Robert Skipper, Jan Slaby, Achim Stephan, and William Wimsatt.

Peggy DesAutels is a philosopher at the University of Dayton who has helped this project along from its inception. She is also my wife and my closest friend. She has contributed to the work at every stage. She even took the time to read and correct the final version—a thankless task for which I thank her. Though sometimes she found my engagement with evolutionary psychology puzzling, she also thought the project was important. She also has contributed in many places to the content, often pressing me to sharpen the point, or at least to make it coherent. I hope in each case I at least met the latter demand. Without her encouragement and interest, I might easily have wandered off into more esoteric concerns.

MIT Press has been very encouraging. July Feldmann did a great deal to improve the work. As an editor, she deserves a great deal of credit.

I have also been fortunate to receive a substantial amount of support from the Taft Faculty committee at the University of Cincinnati over the years. Without that support, I would have had even less time to devote to the project.

Finally, I want to acknowledge a special debt to Thomas Kane, who was formerly a professor within the Department of Biological Sciences at the University of Cincinnati. Tom was a cave biologist, but that underestimates the scope of his knowledge and interest. He studied caves, but he knew enormous amounts about biology beyond those confines. He was comfortable with evolutionary ecology, with population genetics, and with the molecular techniques that inform and shape contemporary evolutionary biology. He loved the fact that he worked in the tradition of great naturalists such as Darwin and Wallace, and he drew from their work as well. I learned much of this from him, both in the field and in the lab. For nearly thirty years, he was a wonderful colleague, and a cherished friend. This book is dedicated to him.

Introduction: Man's Place in Nature

1 Darwin and the Descent of Man

In the final chapter of *On the Origin of Species*, Charles Darwin famously wrote this concerning the implications of his views for human evolution:

In the distant future I see open fields for far more important researches. Psychology will be based on a new foundation, that of the necessary acquirement of each mental power and capacity by gradation. Light will be thrown on the origin of man and his history. (1859, 488)

The last sentence is perhaps the most famous in the *Origin*, since it alone concerns human evolution. It would be easy for us to assume that Darwin meant that *natural selection* would shed light on the "origin of man and his history"; but that assumption would be ill founded. The often-quoted passage comes in the context of a discussion of common descent and the mutability of species, rather than in a context emphasizing the role of natural selection or adaptation. He demonstrates that living things "have much in common, in their chemical composition, their germinal vesicles, their cellular structure, and their laws of growth" (1859, 484) and infers from this that it is likely that all share a common ancestor. He sees that embracing common descent and the mutability of species—what we now call "evolution"—will result in "a considerable revolution in natural history" (ibid.); and he emphasizes that our understanding of classification, or systematics, will need to reflect common descent. Taxonomic classifications, he says, "will come to be, as far as they can be so made, genealogies" (ibid., 486). In this context, natural selection is not even mentioned, although it is alluded to. In the paragraph following the claim, as in the paragraphs preceding it, he returns to the topic of common descent. The contest he enters there is not over the mechanisms of evolution, but the reality of it. Thus it seems clear in the context that he thinks it is, first and foremost, common descent and the mutability of species, rather than natural selection, that will shed light on the "origin of man and his history."

When Darwin finally turns to the topic of human evolution in *The Descent of Man* (1871), his defense should be seen against a historical backdrop in which there was skepticism about evolution as a naturalistic process, as well as skepticism concerning its applicability to human beings. There was also widespread skepticism concerning the role of natural selection. Charles Lyell and Asa Gray, two of Darwin's advocates and friends, had suggested some supernatural impetus was necessary for the evolution of human capacities. That would certainly have offended Darwin's deepest naturalistic sympathies (cf. Richards 1987). Lyell's *The Geological Evidences of the Antiquity of Man* (1863) embraced a deep history for human beings, but Darwin was disappointed to find a less than enthusiastic advocate of evolution. He told Thomas Henry Huxley, one of his closest allies, that he was "fearfully disappointed at Lyell's excessive caution" (in Burkhardt 1983–2001, 11:181). Darwin is careful in the *Descent* first to settle the *question* of descent by focusing on evidence similar to that he had marshaled in the closing chapters of the *Origin*. He carefully took note of the similarity of structure between humans and other mammals, the similarity of embryonic forms, and the presence of "rudimentary" organs (such as male nipples). He observes that the "bearing of the three great classes of facts . . . is unmistakable" (1871, 31). The similarity of structure "between the hand of a man or monkey, the foot of a horse, the flipper of a seal, the wing of a bat &c., is utterly inexplicable" except on the assumption of common descent. Likewise, the vertebral structure we share with apes can be explained by common descent, but not otherwise. Finally, the similarity of early embryonic forms and the presence of rudiments can both be explained on the assumption of common descent, but not on any other assumption. Darwin was particularly inclined to emphasize rudimentary organs—including not only male nipples, but also reduced molars, the appendix, and human tailbones—as things easily explained by common descent, but inexplicable on a doctrine of special creation.

It is important to notice that these appeals do not crucially involve an appeal to natural selection; indeed, they assume that natural selection is *not* the source of the similarity.[1] Darwin was aware that natural selection could—and did in fact—give rise to similarities independently of common descent, and that is exactly why he appealed to similarities of a sort that he did not think were due to natural selection in order to establish common descent. The existence of "rudimentary" structures requires no selection, since the structures are of no significant use to the organism. So they could not be the products of selection. In the ensuing chapters, 2 and 3, Darwin went to great lengths to explore the "mental powers of man," comparing them to those of the "lower animals," carefully including those we fancy to be uniquely human such as language, reason, the moral sentiments, and even self-consciousness. His goal was simple and direct, to show "that there is no fundamental difference between man and the higher mammals in their mental faculties" (1871, 34). The thought he entertained—sometimes using anecdotes we might regard as quaint—was that, for example, curiosity, jealousy, and shame were not peculiarly human, but shared by other primates. He was especially interested in impressing upon the reader that language was not an impossible obstacle for evolution, again suggesting that "monkeys" exhibit language-like skills.

Nevertheless, natural selection certainly had its place in explaining human capacities for Darwin. Again echoing the explanatory scheme of the *Origin*, he noted the existence of individual variations and the tendency of humans to multiply; he concludes "this will inevitably have led to a struggle for existence and to natural selection" (1871, 154). Still the central factor he appeals to in many cases is not natural selection, but sexual selection (see Browne 2002, chap. 9). Much of the two volumes that make up the *Descent* is evidence for the efficacy of sexual rather than natural selection, and Darwin applies that theory in the closing chapters to the human case. Just as peacocks developed tail feathers to enhance their chances of reproduction, and despite any adverse consequences tail feathers might have for survival, so too Darwin thought many of our mental characteristics were favored for their tendency to enhance our reproductive potential. This was what led Darwin to his admission, in chapter 4 of the *Descent*, that he had formerly "probably attributed too much to the action of natural selection or survival of the fittest" in the *Origin* and that he had not "sufficiently considered the existence of many structures which appear to be, as far as we can judge, neither beneficial nor injurious" (1871, 152).[2] Whatever else, in the *Descent*, it is still not natural selection that ends up shedding light "on the origin of man and his history." Common descent is critical, but the cause of common descent is not.

2   The Evolution of Human Values

Darwin was hardly alone in defending an evolutionary account of human abilities. Alfred Russel Wallace—an amazing nineteenth-century naturalist who developed an account of natural selection that was strikingly similar to Darwin's—took up the issue of human evolution in an article in the *Journal of the Anthropological Society* (1864). Like Darwin and Huxley, Wallace defended the idea that humans evolved from an apelike ancestor, but, in opposition to Darwin's views, he maintained that we diversified into races under the influence of natural selection. Thus far, Darwin and Wallace were both Darwinians. As a second theme, Wallace also emphasized the relevance of natural selection to the evolution of our mental profile. Here Wallace was in a sense more a Darwinian than Darwin himself. Natural selection, for Wallace, was supreme.

Neither Darwin nor Wallace was the first to press that evolution should shape our understanding of human psychology. Well before the *Origin*, Herbert Spencer also embraced an evolutionary vision for humans. Spencer was a prominent and imposing figure in Victorian intellectual circles, and was also tremendously influential in American intellectual circles. Spencer was certainly not lacking in intellectual ambition; he was also not lacking in success. His ambitious philosophical program enjoyed the respect of many intellectual stars of the nineteenth century, including John Stuart Mill, Alexander Bain, Joseph Hooker, Charles Lyell, John Tynsall, A. R. Wallace, William James, and of course T. H. Huxley. He was well connected in London society and well regarded, if more than a bit pompous. He in fact was a defender of evolution before Darwin engaged that specific issue in public, contending that evolutionary change was inevitable, not only in organic forms but in social systems as well. He says in *The Principles of Ethics* that his "ultimate purpose, lying behind all proximate purposes" in his intellectual development "has been that of finding for the principles of right and wrong in conduct at large, a scientific basis" (Spencer 1893, 31). His search for a scientific basis for ethics—what he called an "ethics of evolution"—led Spencer through a breathtaking overview of the evolution of the universe, society, psychology, and politics. Spencer's vision was integrated and sustained by his own idiosyncratic evolutionary ideas, which treated psychological, social, biological, and cosmic evolution in nearly the same terms.

The evolutionary theory that led Spencer—inspired mostly by Robert Chambers (1843) and Jean Baptiste de Lamarck (1809)—was one that emphasized diversification. Spencer, like Darwin, found inspiration as well in Milne-Edwards's emphasis on the importance of division of labor in physiology in his *Outlines of Anatomy and Physiology*. Again like Darwin, Spencer found in Malthus an engine to drive social and biological evolution: with the growth of populations, individuals would be forced to accommodate themselves to situations that were increasingly difficult, and the result would be specialization and division of labor. As with political economy, disturbing forces would tend to be corrected for over the longer run if left to themselves. This would happen, in part, because individuals would adapt to their circumstances, and in time these adaptations would tend to be passed on to offspring. He said in *Social Statics* that

The modifications mankind have undergone, and are still undergoing, result from a law underlying the whole organic creation; and provided the human race continues, and the constitution of things remains the same, those modifications must end in completeness. (Spencer 1851, 65)

The purpose of social change, driven by free competition, was a utopian one. The main focus in *Social Statics* was his attack on British social reformers. The perennial issue of reforming the poor laws, he claimed on Malthusian grounds, was an intrusive governmental imposition. The only legitimate function of the state, Spencer held, is the protection of equal rights; the utilitarian defense of the poor laws was an unwarranted excess. Human interference disrupted the natural process. Spencer saw the natural development of human society as a progression toward an ideally classless society in which the natural human sympathies would promote the common good. The mechanism for this evolution is free competition. This is fundamentally *social* evolution.

Even in *Social Statics*, Spencer highlighted his individualism and emphasized our less flattering motives and dispositions. This was balanced by the recognition of the importance of "sympathy" that bonds humans together, as when we care for the suffering or happiness of others. The allusion was to Adam Smith's *The Theory of the Moral Sentiments* (1759), which he thought had laid bare what was at the root of our moral sensibilities. It was supposed to give some traction to the thought that evolution tended to improve the character of the human species, and not just its condition. Spencer maintained much of this general line in his later work, hoping to explain sympathy as the result of evolutionary processes. Like Darwin, he thought it was important to explain the origin of moral feeling in evolutionary terms, though, unlike Darwin, Spencer took this to provide a justification for his social visions. That assessment of their importance was in fact broadly shared by many of his contemporaries.

The next year, Spencer wrote an essay, "The Development Hypothesis" (1852), which, though published anonymously, defended evolution in the *biological* realm by arguing for the influence of circumstance rather than special creation.[3] The framework for Spencer's defense of evolution was both moral and social. Natural and moral laws are essentially identified with each other. So Spencer's evolutionary thought was inherently directional and naturally progressive. Always the dissenter, when Chambers' *Vestiges of the Natural History of Creation* (1843) came under fierce attack from the scientific establishment, Spencer consequently came to be more sympathetic with evolutionary ideas. He argued that an evolutionary account was more credible than any special creationism, and eventually (and somewhat ironically) appealed to Malthus to support a progressive view of life. The pressures of population, Spencer thought, required an increasing specialization and division of labor; this meant that there would be an increase in complexity as a natural result of competition.

As his thinking developed, Spencer increasingly melded the moral and the natural. Robert Richards (1987, 267) says, reasonably, that the moral and social values Spencer held "penetrated to the very root of his scientific considerations, leading him to identify physiological law with moral principles." The ultimate end of evolution, understood progressively, was adaptation, and that was identified as the ultimate moral good. Natural laws became a sanction for moral principles. The utilitarian commitment to the greatest happiness principle, Spencer claimed, was the natural consequence of evolutionary principles. Those who are better adapted to their social environment are those who experience more enjoyment; and conversely, the activities demanded by social life would come to be natural sources of pleasure. The end result of a tortured argument is that the maximization of happiness is not only a natural outcome of evolution, but a moral end as well.

3 Huxley's Attack on Evolutionary Ethics

T. H. Huxley was perhaps the premier public advocate of Darwinism in the nineteenth century. He was doubtless the most visible defender of Darwinism aside from Darwin himself. His works were unabashedly Darwinian, even though he was originally a friend of neither gradualism nor natural selection. He was an uncompromising naturalist and evolutionist. In the 1840s Spencer and Huxley were close friends, intellectual allies, and social comrades. Spencer was the intermediary who, after his return from the voyage on the *Rattlesnake*, facilitated Huxley's entrée into London's intellectual society, a group that included the likes of John Chapman, Marian Evans (George Eliot), G. H. Lewes, Harriet Martineau, and John Stuart Mill (see Desmond 1994, chap. 10). In the mid-1860s Huxley's notorious X-club—an informal group of scientific dissidents—was formed. It included not only Huxley and Spencer, but also Hooker, Tyndall, Busk, Lubbock, and Frankland. All of the X-club had a scientific bent. As Adrian Desmond (1994, 328) says, this was "the new intellectual clerisy, slim and fit after an evolutionary sauna."

Spencer's own evolutionary work was clearly eclipsed by Darwin's *Origin*. Huxley came into his own as a scientist as a defender of Darwin's evolutionary views. Spencer's evolutionary ethics nonetheless continued to play a role in Victorian debates over social policy. He maintained, like Malthus, that policies that supported the poor tended to maintain rather than genuinely assist the poor. Spencer assaulted not only the poor laws, but commercial limits on trade, the national church, war, public education, public health projects, and colonialism. In terms reminiscent of Malthus, Spencer (1851, 322) declared:

the laws of society are of such a nature that minor evils will rectify themselves; that there is in society, as in every other part of creation, that beautiful self-adjusting principle which will keep everything in equilibrium; and moreover, that as the interference of man in external nature destroys that equilibrium, and produces greater evils than those to be remedied, so the attempt to regulate all the actions of a people by legislation will entail little else but misery and confusion.

There could be no natural injustice addressed by the poor laws since the return was exactly what was deserved. The solution was to abolish the poor laws. He concluded, in reflecting on the implications for the broader population, "If they are sufficiently complete to live, they *do* live, and it is well they should live. If they are not sufficiently complete to live, they die, and it is well they should die" (1851, 414–415).

Huxley, by contrast, had more humane values. He had long been committed to state-sponsored education (Huxley 1871; see Desmond 1994), thinking that anything less simply guaranteed the status quo. By the late 1880s Spencer and Huxley were clashing openly over land reform, with Huxley sharply critical of Spencer's opposition to state involvement, as it collided with Huxley's concern for educational reform. As was often the case with Huxley, the split was not amicable. In "The Struggle for Existence in Human Society," Huxley (1888, 199) is blunt in dismissing Spencer's vision:

it is an error to imagine that evolution signifies a constant tendency to increased perfection. That process undoubtedly involves a constant remodeling of the organism in adaptation to new conditions; but it depends on the nature of those conditions whether the direction of the modifications effected shall be upward or downward. Retrogressive is as practicable as progressive metamorphosis.

Huxley's point was not just about the biological realm, but also about the social. In the wake of his daughter's death, he came to think that the key to moral progress lay in resisting rather than acquiescing to suffering. He says, in "The Struggle for Existence in Human Society," that it is only the savage that "fights out the struggle for existence to the bitter end" (1888, 198). Civilized people respond by resisting. Huxley expected a confrontation with Spencer, who responded by agreeing with Huxley's observation that nature observed no moral course.

Their differences came to a head over Huxley's *Romanes* lecture at Oxford, "Evolution and Ethics" (1893). It was an unrelenting attack on Spencer's identification of moral with evolutionary progress. Huxley (1893, 79) says, with what seems more than a little irony, that Spencer "adduces a number of more or less interesting facts and more or less sound arguments" concerning the origin of the moral sentiments by natural means. Mostly he regarded the arguments and the facts as less rather than more sound. He pointed out crucially that the "immoral sentiments" likewise have a natural origin. Evolution is as indifferent as is the Victorian God to human suffering. "The thief and the murderer," he says, "follow nature just as much as the philanthropist" (ibid., 80). In connection with the Darwinian principle of "survival of the fittest," Huxley observes that this too is prone to causing confusion, and that this is part and parcel of Spencer's evolutionary speculations. Huxley complains that Spencer assumes that since the struggle for existence leads to increased complexity, humans too must embrace the struggle for existence as the means to bettering human existence. Huxley (1893, 60) observed to the contrary that nature is indifferent to human suffering or pleasure, that "grief and evil fall, like the rain, upon both the just and the unjust." In the case of pain and suffering, he says, this "baleful product of evolution increases in quantity and in intensity, with advancing grades of animal organization, until it attains its highest level in man" (ibid., 51). Even with regard to the human sentiments, Huxley conceded that the moral sentiments, such as sympathy, were a natural element of humans and the result of evolution—but so were the immoral sentiments, such as revenge and lust. Evolution was indifferent to moral character. Huxley took the response one step further. The *moral* response was to *resist* the tendencies of nature:

the practice of that which is ethically best—what we call goodness or virtue—involves a course of conduct which, in all respects, is opposed to that which leads to success in the cosmic struggle for existence. In place of ruthless self-assertion it demands self-restraint; in place of thrusting aside, or treading down, all competitors, it requires that the individual shall not merely respect, but shall help his fellows; its influence is directed, not so much to the survival of the fittest, as to the fitting of as many as possible to survive. (Ibid., 81–82)

Where natural selection might favor "immoral instincts," such as ruthless self-assertion, morality requires self-restraint. Morality, Huxley says (ibid., 82), "repudiates the gladiatorial theory of existence." The point of ethics, Huxley thought (ibid., 52), was to find the "sanction" of morality—to tell us "what is right action and why it is so." It was certainly not to sanctify greed. The appeal to natural processes, he contended, failed precisely because it could discern no line between the moral and the immoral, the just and the unjust.

4 Evolutionary Naturalism

Huxley and Spencer were in many ways natural allies. Both were certainly naturalists, especially in opposing religion. Both were certainly evolutionists and anxious to apply evolutionary principles to human beings. Huxley's *Man's*

Place in Nature (1863) was, after all, a manifesto supporting the physical and psychical unity of humans with other animals. He was willing not only to concede but to insist that the moral sentiments had their origin in evolution just as do all natural phenomena. But the two men certainly parted company over aspects of their social agendas and over their explanations of human capacities. Spencer's harsh vision was strikingly different from Huxley's own progressive commitment to the working poor and educational reform. Constitutionally, Huxley was averse to speculation, emphasizing the human reality rather than the social ideal. Both were indeed naturalists, but Huxley's was a leaner naturalism.

Evolutionary psychology, too, offers us a form of evolutionary naturalism, committed to the idea that natural processes are responsible for the evolution of human capacities. These are commitments I share. Human beings, like other organisms, are the products of evolution. Our psychological capacities are evolved traits as much as our gait, dentition, or posture. Furthermore, Human beings, like other organisms, exhibit traits that are the products of natural selection. Our psychological capacities are subject to natural selection as much, again, as our gait, dentition, or posture. In this minimalist sense, there can be no reasonable quarrel with evolutionary psychology. Creationists and advocates of "rational design theory" might quarrel with an evolutionary vision, but these quarrels fly in the face of accumulated knowledge. It is incontrovertible that evolution is real. It is a theory. It is also a fact. It is worth neither defending nor disputing the fact that we are the products of evolution any more than it is worth disputing that our bodies are composed of cells or that the Earth circles about the Sun. It is also clear that natural selection has had a role in shaping life on Earth. It is worth neither defending nor disputing the fact that humans are the products of natural selection any more than it is worth disputing that we are mortal or that we are bound to the Earth by gravity. Those who would argue with evolution should look elsewhere for comfort.[4] The issues I will raise here are issues that fit comfortably within evolutionary theory.

If evolutionary psychology settled for such uncontroversial conclusions, it would in turn be uncontroversial. At least, it would not be a subject of *scientific* controversy, though it is subject to controversies over science. As I've said, I'm engaged primarily in the former. Evolutionary psychology is certainly controversial and ambitious. In the first chapter, I'll spend some time describing the ambitions that are characteristic of evolutionary psychology. In subsequent chapters, I'll express considerable skepticism concerning these evolutionary ambitions.

To give a flavor of the kind of issues I'll raise in ensuing chapters, consider the work from Donald Symons. His work is focused on human sexual preference—what males and females prefer in sexual partners. I think this is work that is characteristic of work within evolutionary psychology; indeed, it is more thoughtful and reflective than much of it. Symons (1992, 141) thinks the key question is what he calls the "adaptationist question": Was a particular trait, or behavior, "designed by selection" to serve some function? To that, of course, we should add the further question, "What function did it serve?" It is hard to overstress the importance of history, and that what is being evaluated here is a historical claim. Symons, for example, has offered us a striking array of evidence concerning sexual preferences (see, e.g., Symons 1979, 1992). The basic picture is easy to understand. Human males are more attracted to youthful women. Human females are more attracted to high-status men. Symons recognizes that claiming that sexual preference is an adaptation is to advance a historical claim. It is a claim about the evolutionary history of sexual preferences, and that concerns our behavior in ancestral environments. It is not fundamentally a claim about current differences. He claims nonetheless that there is a wide array of evidence available to the evolutionary psychologist:

In evaluating the hypothesis that human males evolved specialized female-nubility-preferring mechanisms, here are some of the kinds of data that might prove to be relevant: observations of human behavior in public places, literary works (particularly the classics, which have passed the tests of time and translation), questionnaire results, the ethnographic record, measurement of the strength of penile erection in response to photographs of women of various ages, analyses of the effects of cosmetics, observations in brothels, the effects of specific brain lesions on sexual preferences, skin magazines and discoveries in neuropsychology. (Symons 1992, 144)

Any of these things *might* prove to be relevant, of course, so long as "relevance" is a sufficiently weak relationship. A look at pornographic magazines at least suggests that those who regularly buy them like youthful women. That doesn't tell us much about those who do *not* buy them (and I am skeptical that most males do so regularly). A look at scuba diving magazines, after all, suggests that those who buy them are interested in scuba diving. It doesn't tell us much about those who do not buy them (certainly, most do not). The focus on pornography offers some probabilistic support to the hypothesis that men like youthful women. The focus on scuba magazines offers some probabilistic support to the hypothesis that people like diving. Presumably something stronger than simple relevance is intended.[5] Any evidence *might* prove relevant, if relevance is construed in a liberal way. Likewise much evidence might, for example, disconfirm the claim that there are such preferences, and thereby disconfirm the claim that there are such "mechanisms." For the moment, at least, let's generously assume the evidence favors the claim that there are such preferences among human males. Let's even suppose that it supports the claim that these preferences are relatively stable across cultures. What would follow? Not much. More specifically, such evidence would not support the view that human males evolved such preferences. The evidence for that might be taken in two ways, but in either case, the support would be very weak indeed. It might be a claim about the evolution of *male* as opposed to *female* preferences. This is a hypothesis partly about differences between males and females. Perhaps, say, observations in brothels would support some conclusions about male preferences. They would tell us little about sexual differences, or the forces that shaped them. Studies of penile erection are likely to be equally uninformative on this question.

I suppose Symons is mainly interested in the more general evolutionary question concerning sexual attraction. It is important, though, that if we were offered a comparable claim concerning the evolution of male and female preferences in birds, we would expect to see evidence relevant to the differences in preference and not merely to overall or average preferences. Evolution proceeds on variation rather than averages. So let's suppose we found good evidence that there were such differences in preferences. Would that support the conclusion that the differences are "evolved"? Inevitably, in the minimalist sense. We have them. We've granted there are differences. They came from somewhere. Our ancestors are the only candidates. No serious evolutionist is interested in such claims, of course. Assuming the "differences" are robust and real, and not attributable to developmental differences, we would need to explain them somehow.

What evolutionary *explanation* could we offer? Evolutionary psychologists tend to assume that the only explanations will be in terms of adaptation. Real differences must be the products of natural selection. This is where the "constraints" on adaptation explanations come into play. I'll explore these in more detail later. For now, an illustration should suffice. Bipedalism is a characteristic of humans. So is a large brain case. Both are evolved. But the cases are importantly different. Very roughly, bipedalism is certainly not a specifically human adaptation. Our hominid ancestors were also bipeds. It would be a mistake to explain bipedalism as an adaptation to *our* ancestral environment. A large brain case is specifically human. It is characteristic of the genus we belong to. It evolved within that lineage. Let's return to Symons, given the broader evolutionary context. None of the issues Symons introduces address the fundamental evolutionary questions. He offers some evidence concerning the preferences that are present. He assumes that humans would have benefited from such preferences. Perhaps they would have. Perhaps not. Knowing that would depend on knowing the variation in ancestral populations. In any case, nothing suggests these preferences are specifically human features. Let's suppose they are specifically human. To show that the preferences are adaptations, even this would not be enough. Grant that there are such preferences. Grant that they are common across cultures. Grant further that they are specifically human features. Even grant that they evolved within humans. Would it follow that they are adaptations? Again, it would not. To show that, we would need to show that they were the products of natural selection. For that, we would need evidence concerning variation in ancestral populations. We would need evidence concerning their heritability. And if we wanted a full explanation of their presence, we would need evidence concerning the advantage they offered to our ancestors. The evidence Symons would have us appeal to is simply silent on such matters. It is equally silent on what would cause any supposed differences in fitness.

What I think we should demand of evolutionary psychology is evidence specifically supporting the evolutionary claims they offer. Of course, it is reasonable to expect their claims to pass muster as *psychology*; but my focus will be on the *evolutionary* credentials on offer. I share their evolutionary vision. As Huxley challenged the evolutionary credentials of Spencer's theory, I will mount an evolutionary challenge to evolutionary psychology. As Spencer and Huxley shared an evolutionary vision, I share with evolutionary psychologists an evolutionary perspective. As Spencer and Huxley differed over what this perspective warrants, I will depart from evolutionary psychologists. As Huxley viewed Spencer's evolutionary ethics with suspicion, I view evolutionary psychology with suspicion. As Huxley viewed Spencer's theory as more speculation than science, I view evolutionary psychology as more speculation than science. The conclusion I urge is, accordingly, skeptical. Speculation is just that: speculation. We should regard it as such. It does not warrant our acceptance. Evolutionary psychology as currently practiced is often speculation disguised as results. We should regard it as such.

1 The Ambitions of Evolutionary Psychology

1 Darwin's Gift

Evolutionary psychology, as I have said, does not suffer from lack of ambition. Neither does it shy away from controversy. The acknowledgment of our place among animals, subject to natural selection, is integral to its vision, but far too modest adequately to comprehend its ambitions. At its roots, evolutionary psychology is fundamentally a psychological program, geared to the reform of psychology as a science. As it has been developed, evolutionary psychology also embraces an aggressive biological program. On the more ambitious program characteristic of evolutionary psychology, psychological processes are adaptations, not to present circumstance, but to our ancestral environment. The fact that they are adaptations, in turn, is supposed to explain and ground our psychological capacities. Leda Cosmides and John Tooby, two of the most prominent figures in the field, offer this description of the agenda and put it in a compelling evolutionary perspective:

> The human mind is the most complex natural phenomenon humans have yet encountered, and Darwin's gift to those who wish to understand it is a knowledge of the process that created it and gave it its distinctive organization: evolution. Because we know that the human mind is the product of the evolutionary process, we know something vitally illuminating: that, aside from those properties acquired by chance, the mind consists of a set of adaptations, designed to solve the long-standing adaptive problems humans encountered as hunter-gatherers. (Cosmides and Tooby 1992, 163)

The mind is a dauntingly complex phenomenon. Features with complex functional designs must after all have evolved, and in order to evolve, they must have provided a substantial advantage to our ancestors in virtue of their design. These advantages explain the current structure and the prevalence of these features. It is not enough that the mind evolved or even that the mind is subject to natural selection. It is not enough that humans evolved or that our evolution is subject to natural selection. The complex functional designs we observe must have been selected *for* among our ancestors: these features obviously came to be as the products of evolution. Cosmides and Tooby hold that they not only evolved but were selected for, that they were specifically favored by natural selection. Otherwise they would not exist. This is what Cosmides and Tooby call "Darwin's gift." The key thought is that natural selection is required to explain the capacities we exhibit. Natural selection is required to explain the complexities of judgment, thought, perception, emotion, and action. Natural selection is required to explain who and what we are. If Darwin is right, they think, then natural selection must also suffice to explain who and what we are. The advantage offered by our cognitive organization must explain its presence. So, too, our perceptual abilities, and our emotions, must reflect our evolutionary history and the prevalence of natural selection. The mind is no different than any other complex feature. The mind is an adaptation. That, at least, is "Darwin's gift." It lies at the heart of the evolutionary agenda for evolutionary psychology. Its advocates intend it to be what Thomas Kuhn thinks of as a paradigm shift, a dramatic reorientation in the way we conceive human behavior.

What evolutionary psychology offers is ambitious in another way. Evolutionary psychology supplies a comprehensive agenda that applies to a broad range of characteristically human behaviors. It provides a broad characterization of human behavior, together with an explanation for it: We are aggressive in defending family and territory; indeed, we humans are remarkable for our ferocity and our willingness to engage in gratuitous violence. Among other things, we are likely responsible for the extinction of the striking megafauna characteristic of the Americas toward the end of the last Ice Age, in what is called "Pleistocene overkill"; and we certainly deserve credit for the demise of less dramatic forms such as the dodo. And of course, we happily exterminate our own kind. We have complex sexual relations. There are differences between males and females in terms of what we value and how we behave. We are often afraid of strangers and of heights. There are conflicts between children and their parents, and there are differences between parents. On the more positive side, we engage in complex play. We engage in a variety of cooperative behaviors and have lasting friendships; we form lasting personal bonds. We are at least as curious as we are violent.

Much of the work in evolutionary psychology is devoted to documenting such complex patterns and explaining them. The very patterns of behavior themselves are, of course, matters of controversy. David J. Buller's *Adapting Minds* (2005) explores these controversies in considerable detail, focusing especially on what evolutionary psychologists say about mating, marriage, and parenthood. He shows, to my mind convincingly, that there are alternative explanations of the behaviors we observe, and that in some cases the predictions concerning what we observe are themselves problematic. The possibility of an alternative does not show that it is true, of course, and Buller offers some support for his preferred views. The explanations inherit these uncertainties, whatever they may be. Though the psychological evidence will not be central to my discussion, we'll see that the uncertainties concerning what needs to be explained are considerable. The explanations are more problematic still.

Assuming the patterns are real, though, evolutionary psychology offers explanations for such human tendencies. Some human violence is evidently geared to the acquisition of resources. It is tied to sexual rivalry, to power, to wealth, and to status. In his famous studies of the Yanomamö, the anthropologist Napoleon Chagnon (1983) illustrates these tendencies in great detail. Men will raid for food or for women, and they engage in ritualistic fights for status. Evolutionary psychology takes mate preferences, accordingly, to be evolved rather than socially derived responses. It is a social accident if a Yanomamö male uses a steel axe on a rival, since the axe was introduced, but the aggression itself is natural. Thus the tendency toward aggression is to be explained in evolutionary terms, even if the specific form it takes is socially shaped. In like manner, the psychologist David Buss traces the differences between the sexes finally to differences in investment between the sexes: in terms of mate preference, Buss tells us females favor mates who are dependable, stable, and high status, whereas men prefer mates with youth and health (see Buss 1994, 1999). How status is measured may be socially variable, but a preoccupation with status is not; after all, it has evolutionary consequences. A female with a high-status mate will likely have offspring that have the enhanced reproductive potential that supposedly accompanies high status. Matt Ridley (1993, 118) captures the view:

Wherever you look, from tribal aborigines to Victorian Englishmen, high-status males have had—and mostly still do have—more children than low-status ones. And the social status of males is very much inherited, or rather passed on from parent to child.

These "facts" are controversial. Many believe that the economically disadvantaged reproduce at a higher rate; and in any case, increased education levels tend to reduce family size. Many also believe that increased rates of reproduction among the poor pose a threat to social well-being. These are controversies I don't want to engage in here. The picture is clear: social status is supposed to lead to enhanced reproductive potential, at least among our ancestors (even if not Victorian Englishmen), and that is supposed to explain our current perceptions of and preoccupations with status.

On this view, at least some human fears (but not all) are given explanations in evolutionary terms. So a fear of snakes or spiders, like our fear of strangers or of heights, supposedly serves to protect us from dangers. Having observed that snakes and spiders are *always* scary, and not only to humans but to other primates, Steven Pinker (1997, 386) says "The common thread is obvious. These are the situations that put our evolutionary ancestors in danger. Spiders and snakes are often venomous, especially in Africa. . . . Fear is the emotion that motivated our ancestors to cope with the dangers they were likely to face" (cf. Nesse 1990). This is a curious view, actually. Spiders offer very little risk to humans, aside from annoyance. Most are not even venomous. There are perhaps eight species of black widow, one of the Sydney funnel web, six cases of the brown recluse in North and South America, and one of the red banana spider in Latin America. These do present varying amounts of risk to humans. They are not ancestrally in Africa, our continent of origin. Given that there are over 37,000 known species of spiders, that's a small percentage. The risk from spiders is exaggerated. The "fact" that they are "always scary" and the explaination of this fact in terms of the threat they posed to our ancestors is nonetheless one piece of the lore of evolutionary psychology.[1] Likewise, snakes have a reputation among evolutionary psychologists that is hardly deserved. In Africa, some are truly dangerous, but by no means most. About one quarter of the species in Uganda pose a threat to humans, though there is geographic variability. It's only in Australia—hardly our point of origin—that the majority of snakes are venomous. Any case for an evolved fear of snakes would need to be based on the threat from a minority. In this case too, the threat seems exaggerated. There is a good deal of mythology in the anecdotes we are offered. It is not altogether clear how the mythology gets established, but it is often repeated, with scant evidence. I'll reinforce this moral in what follows.

Anecdotes are often reinforced by powerful theory. The theory of parent–offspring conflict developed by Robert Trivers (1974) is used to explain the differing "interests" of parents and their offspring with respect to, say, the use of resources. This in turn is supposed to undergird the conflict between parent and offspring. The explanation of cooperation, and lasting friendships, comes from what is called "reciprocal altruism," a model telling us that coop-eration can be favored when there is some mutual benefit to be derived (Trivers 1971; Axelrod 1984; Axelrod and Hamilton 1981).

This is only a sampling of the psychological and social domain claimed by evolutionary psychology, but it does illustrate the challenging range of behav-iors it aims to explain. One common factor among these explanations is that they explain the patterns *as* adaptations; that is, the patterns are explained as the products of natural selection acting over generations, molding behavior to the demands of survival and reproduction in our ancestors. Cosmides and Tooby (1994, 530) put it this way:

Natural selection shapes domain-specific mechanisms so that their structure meshes with the evolutionarily stable features of their particular problem domains. Understanding the evolutionarily stable feature of problem domains—and what selection favored as a solution under ancestral conditions—illuminates the design of cognitive specializations.

How do we know that these traits are adaptations? Sometimes this is simply assumed. It is actually a serious issue, one that occupies evolutionary biologists. The most straightforward argument for a focus on adaptation as the engine of evolution starts with the complexity of the features. As I've already said, this is where Cosmides and Tooby begin. The point is not merely that complex features are evolved. All our features have evolved. The simplest and most fundamental features are inherited. They may be relatively constant; they may change; but all have evolved. All depend on our evolutionary heritage, just as they depend on our development. However simple or complex they might be, they can be given an evolutionary explanation. At an early stage in development, for example, the human fetus has a characteristic radial cleavage (the result looks like a spiral staircase) that we share with other vertebrates, but differs from the pattern we see in arthropods (which looks more like stacked spheres). We share this developmental pattern with vertebrates because it was inherited. It evolved as part of the pattern. Adaptation, however, requires more than this. This is where the appeal to complexity enters, though our development is certainly astonishingly complex.

The roots of the appeal to complexity, perhaps paradoxically, lie in natural theology. In the late eighteenth century, William Paley, deacon of natural theology, argued that complexity demanded an intelligent designer. Darwin had studied Paley while he was a student at Cambridge. One of Darwin's key insights, put to great effect in the *Origin*, was that it is possible to explain design without intelligence. He embraced the thought that complex features present a special problem for evolution. Darwin (1859, 3) wrote this in the introduction to *On the Origin of Species*, doubtless with Paley in mind:

In considering the Origin of Species, it is quite conceivable that a naturalist, reflecting on the mutual affinities of organic beings, on their embryological relations, their geographical distribution, geological succession, and other such facts, might come to the conclusion that each species had not been independently created, but had descended, like varieties, from other species. Nevertheless, such a conclusion, even if well founded, would be unsatisfactory, until it could be shown how the innumerable species inhabiting this world would have been modified, so as to acquire that perfection of structure and coadaptation which most justly excites our admiration.

In contemporary terminology, similarities, development, distribution, and succession support evolution; yet complexity in structure and adaptation require an explanation as well. Modern evolutionists often follow Darwin. The presence of complex features, those exhibiting "perfection of structure and adaptation," we are told, must be the consequence of evolution by natural selection. Evolutionary psychologists are enthusiastic in endorsing the connection. Pinker embraces the line, saying "Natural selection has a special place in science because it alone explains what makes life special. Life fascinates us because of its *adaptive complexity* or *complex design*" (Pinker 1997, 155; cf. Tooby and Cosmides 1992, 49ff.; Pinker and Bloom 1992; Grantham and Nichols 1999). I think of this as *Dawkins' gambit*. In *The Extended Phenotype*, the British biologist Richard Dawkins (1982, 43) says "if we see an animal with a complex organ, or a complex and time-consuming behavior pattern, we would seem to be on strong grounds in guessing that it must have been put together by natural selection. . . . The working hypothesis that they must have a Darwinian survival value is overwhelmingly strong." Dawkins admits that this assumption can be overturned, but complexity makes it the best "working hypothesis." The idea is intended by Dawkins, no less than Cosmides and Tooby, to be straightforwardly Darwinian: natural selection is the *only* available explanation for the evolution or presence of complex functional designs. Other evolutionary factors such as mutation or drift, by contrast, will not tend to lead systematically to such complex features, and if there are constraints on the evolutionary process, those would tend to reduce rather than facilitate adaptation. In this Darwinian sense, these alternative evolutionary mechanisms can be regarded as "chance" factors, uncorrelated with evolutionary advantage. Tooby and Cosmides (1992, 57) say "It would be a coincidence of miraculous degree if a series of these function-blind events, brought about by drift, by-products, hitchhiking, and so on, just happened to throw together a structure as complexly and interdependently functional as an eye."

In an even more dramatic fashion, Buss acknowledges that although some features evolve as "by-products" or "spandrels," and although some features are prevalent by chance, it is natural selection that occupies center stage for evolutionary biology. Here is a representative passage from Buss (1999, 39):

Despite scientific quibbles about the relative size of the three categories of evolutionary products, all evolutionary scientists agree on one fundamental point: adaptations are the primary product of evolution by natural selection. . . . Those characteristics that pass through the selective sieve generation after generation for hundreds, thousands, and even millions of years, are those that helped to solve the problems of survival and reproduction.

He concludes that "the core of all animal natures, including humans, consists of a large collection of adaptations" (ibid.). Evolutionary psychologists need not deny that there are other evolutionary factors at work; they need not deny the workings of "chance." But they do characteristically focus on natural selection. Evidently, this is because they are convinced that in doing psychology we are faced with features so complex that they demand explanation as adaptations.[2]

Dawkins' gambit is hardly uncontroversial. There are several available explanations for adaptive complexity, and not all require that the traits be adaptations. Developmentalist alternatives are among them. The most striking recent additions, which derive from developmentalist traditions, are those that appeal to the emergence of complexity. Stuart Kauffman (1993), among others, has been instrumental in developing a science of complexity. Kauffman claims that the problem for twenty-first-century science is to explain "organized complexity," including ecosystems, communities, organisms, genetic regulatory systems, and neural systems. His exploration of the "origins of order" emphasizes that across disparate domains simple general principles suggest that there is a natural and spontaneous order in complex systems, apart from, and prior to, adaptation. Robert E. Page and Sandra D. Mitchell (1991) illustrate the point using colonial insects: they exhibit a complex social structure, which, Page and Mitchell suggest, depends on the dynamics of self-organization rather than adaptation. It turns out, in point of fact, not to be difficult to generate the complex caste structure of social insects (see also Mitchell 2003). These explanations are controversial in a number of ways (see Burian and Richardson 1991, and Richardson 2001b, for critical assessments). I use them primarily to illustrate that there are alternatives to Dawkin's gambit. Complexity can have many sources.

The central programmatic goal of evolutionary psychology, correspondingly, is to provide evolutionary explanations of our natural psychological capacities in terms of natural selection (see, e.g., Grantham and Nichols 1999; Ridley 1993). This is not just a matter of arguing that some psychological feature is an adaptation; it requires knowing what that feature is an adaptation *for*. Even if we bought Dawkins' gambit, accepting that complex features must be adaptations, it is a far more difficult task to explain them in terms of natural selection—that is, to show *what* they are adaptations *for*. This is exactly what evolutionary psychology requires. To take the most prominent examples, Cosmides and Tooby claim that human reasoning consists of a set of mechanisms organized around social exchange. They describe the program of research this way:

According to the evolutionary psychological approach to social cognition . . . the mind should contain organized systems of inference that are specialized for solving various families of problems, such as social exchange, threat, coalitional relations and mate choice. . . . Each cognitive specialization is expected to contain design features targeted to mesh with the recurrent structure of its characteristic problem type, as encountered under Pleistocene conditions. Consequently, one expects cognitive adaptations specialized for reasoning about social exchange to have some design features that are particular and appropriate for social exchange, but that are not activated by or applied to other content domains. (1992, 166)

According to Cosmides and Tooby, the evolutionary function of human reasoning involves facilitating and monitoring social exchange and social relations. Human reasoning then would be a "cognitive adaptation" to social conditions encountered by our evolutionary ancestors, established and maintained by natural selection. In this case, the social context of hominid life shapes our behavior and ways of thinking. The initial goal of evolutionary psychology, thus, is explaining psychological processes as biological adaptations to Pleistocene conditions, adaptations that have been shaped and maintained by natural selection. It does not follow, of course, that human psychology is adapted to our current conditions—conditions of increased crowding, dissociation from extended family, larger social groups, and overwhelming amounts of information that travel, thanks to computers, faster than the speed of cars, or even the speed of sound. Human reasoning may now be less than optimal, but how we think today is supposed to be explained by our history. Once that history is exposed, we may even find that we can redesign our environment to better suit our natural ways of thinking (see, e.g., Gigerenzer 1998). According to evolutionary psychologists, this is true not only for social cognition but for a wide array of psychological mechanisms and social behaviors. Similar explanations, as I have said, have been offered by evolutionary psychologists for family structure, parental care, marital jealousy, sex roles, sexual preferences, familial affection, personality, and the moral sentiments, to mention just a few. Doubtless more will follow.

2 Darwinian Algorithms

The evolutionary explanations offered by evolutionary psychologists are a means to an end, where the end is the reform of psychology. I've already noted that this move has a venerable heritage, with Herbert Spencer as one who offered a reformed psychology, and William James as one who would have functionalist psychology conform to evolutionary visions. Most of the recent advocates of evolutionary psychology are themselves psychologists or anthropologists rather than biologists. The problem many of these psychologists see in their own field is a kind of malaise following on the lack of a definite vision. James, Angell, and Dewey saw a similar malaise at the turn of the twentieth century and offered a similar evolutionary prescription. Evolutionary biology offers a vision that, they claimed, could transform psychological research. Evolutionary psychologists echo the thoughts of nineteenth-century function-alists in psychology. The key synthesis of psychology and biology that evo-lutionary psychology offers is meant in the end to reform psychological research, not to reform our conception of human biology. The contributions on offer come at a number of distinctive levels; but it is, fundamentally, a psy-chological program.

Tooby and Cosmides are fundamentally interested in the potential of evo-lutionary biology for grounding work in psychology and the social sciences. "Modern biology," they say, "constitutes, in effect, an 'organism design theory'" (1992, 53). It can be used as a lever both to undercut the "standard" models in social science and to "guide the construction" of an improved psy-chology. Donald Symons (1992) similarly tells us that Chagnon's work on the use of terms for kinship among the Yanomamö was inspired by a selectionist vision. No doubt, this is so. Kinship is here a social matter, and Chagnon sees it in these terms. The particular category to which an individual is assigned affects, among other things, their eligibility for marriage. Chagnon noticed that people do not simply accept the existing classification if they can benefit by changing it. So, for example, Chagnon (1998) describes one Yanomamö man who called a young woman by a name that enhanced her sexual eligibility for his son. Symons takes this to be a defiance of social imperative. Chagnon's conclusions are more modest than Symons's. Chagnon does not offer here any-thing like an "organism design theory." Symons's point is nonetheless surely correct: there is a contribution from biological to social theory, in this case to anthropology.

There are other obvious contributions evolution could make to the under-standing of human psychology. If biology offers an "organism design theory," evolutionary psychology could, with little overstatement, be said to offer a science of "human nature." It begins with a description of our adaptation to ancestral environments and moves to explaining our current capacities, includ-ing our psychological capacities. To take one example, many evolutionary psy-chologists are prepared to accept that the modularity of mind is a natural consequence of evolution (see, e.g., Fodor 1985; Baron-Cohen 1995). The thought is that that the mind is an amalgam of special-purpose mechanisms, rather than a general purpose machine. So as phrenologists portrayed the mind as a mosaic of faculties two centuries ago, we still find the mind portrayed as a mosaic of relatively independent modules, though of course the emphasis on the shape of the skull has been replaced with PET and fMRI, allowing us to glimpse the functioning of the brain in action.

That a modular organization would have facilitated the evolution of mind is often taken as obvious. Symons (1992) dismisses any alternative out of hand with the observation that there is no such thing as a general problem solver, since there is no such thing as a general problem. Cosmides and Tooby are equally insistent, but not quite as dogmatic. Here is one passage:

> A basic engineering principle is that the same machine is rarely capable of solving two different problems equally well. . . . Our body is divided into organs such as the heart and the liver, for exactly this principle. Pumping blood throughout the body and detoxifying poisons are two very different problems. Consequently, the body has a different machine for solving each problem. (Tooby and Cosmides 1992, 80)

Functional specialization is good engineering design. So, similarly, if the mind is a machine selected for its ability to cope with environmental demands, then it should also "consist of a large number of circuits that are *functionally specialized*" (ibid.). One key thought Cosmides and Tooby want to enforce is that if we assume an evolutionary perspective, then the mind must consist of a set of specialized systems, each "designed" to solve some relatively specific family of problems, such as problems of social exchange, mate choice, threat, and the like. The alternative is to assume that our reasoning consists in the application of relatively general procedures that are not content specific. General procedures are unlikely, they claim, to yield adequate solutions. There are several distinct questions one might want to answer. First, *is* our cognitive organization modular? This is not a question to which I claim to know the answer.[3] Second, if it is modular, what sort of content specialization should we expect? Third, what are the implications of adaptive thinking for such questions? Even at the first level, we should be cautious. The idea that there is modular organization in the brain, with specialized and relatively independent subsystems, is controversial (see, e.g., Uttal 2001; Bechtel and Richardson 1993). This is an issue that would take me too far afield. Evolutionary psychology likely has more implications for the second question than the first. If we know the evolutionary "problems" our ancestors faced, then we should be able to project the kinds of specialized "problem-solving" abilities they might have had. They surely did not need the capacity to navigate Los Angeles freeways, though no doubt the ability to coordinate activities among themselves was desirable. As Aristotle would have acknowledged, we are not squash players by nature, though we are naturally political animals. We did not need the capacity to navigate at the speed of sound, though no doubt sensitivity to flying objects was desirable. We did not need the capacity to play chess, though spatial problem solving is doubtless important. The third question—what are the implications of adaptive thinking—is the one that will occupy us in many ways. That is one reason this is a philosophical piece. Nonetheless, the questions are not independent. The implications depend, like the Devil, on the details.

We can begin with the connection between psychological data and the interpretation of it by evolutionary psychologists. The first commitment aims to tie the psychological data to an evolutionary interpretation relatively directly. So we are offered a contrast between two, or perhaps three, pictures. One is what Tooby and Cosmides call the "Standard Social Science Model," according to which humans come equipped with only general-purpose rules for learning. The other extreme is the "Evolutionary Psychological Model," according to which, in addition to whatever general-purpose rules there might be, there are also special-purpose, content-specific rules for learning. The intermediate position would naturally accommodate both special-purpose and general-purpose mechanims (see, e.g., Fodor 1985). Cosmides and Tooby suggest, more specifically, that there are specialized cognitive procedures—Darwinian algorithms—for social exchange. Here is one way they tie the psychological data to their biological interpretation. If there were *only* general-purpose learning rules, they reason, then all the specific content would be attributable to "cultural" influences. The rules, if wholly general, could provide no specific content. The social environments into which we are born would provide the needed "content" for social judgment. If, on the other hand, there are specialized rules for social exchange, these could form "the building blocks out of which cultures themselves are manufactured," and these rules would constitute a kind of "architecture of the human mind" undergirding a social exchange psychology. Every human being would come with the same basic cognitive equipment; the traditions, rituals, and institutions characteristic of human life at most would supply the "specifics" (Cosmides and Tooby 1992, 208). We might be geared cognitively to discern cheaters, for example, though the specific opportunities for cheating vary substantially.

John Alcock is a distinguished ethologist and a student of Ernst Mayr. In *The Triumph of Sociobology* (2001), he offers an extended attack on the "social science" model. Feminists, sociologists, and other academics, on his view, evidently resist evolutionary explanations in favor of "proximate" ones that focus on more immediate causes, including social and psychological influences. Alcock points out, reasonably, that the outcomes are not *merely* the effects of culture, though none of his opponents would obviously disagree on that specific point, even if they do often emphasize the significance of culture. They need not deny a role for biology just because they allow a role for culture.

(Neither need one deny a role for culture in order to allow a role for biology.) Alcock (2001, 140) goes on to recount the data from Daly and Buss that there are similar patterns of attractiveness of women across widely different cultures and that that supports the idea that "standards of beauty adopted by males encourage men to pursue fertile women." This is a biologically uniform standard, according to Alcock. Again, I'm less interested at the moment in the (truth of the) cultural claim than in the reasoning behind it. Like many evolutionary psychologists, Alcock infers from the supposed universality of a pattern of preference that there is a biologically grounded pattern, namely a Darwinian algorithm.

This much is meant to mark out the differences between their own psychological model and the "social science" alternative. Cosmides and Tooby then move on to draw out two conclusions, both of which they take to favor evolutionary psychology. They say, first, that if there is a "universal evolved architecture of the human mind," then there should be some features of social exchange that are common across individuals and across cultures. And, they tell us, this is exactly what we find. It is also what we supposedly find in preferences concerning the standards of beauty. The implication is, supposedly, that we would *not* find this if there were only generalized learning mechanisms. This latter implication is plainly not true. Even if there were only generalized learning rules, we would expect *some* features to be common across cultures and across individuals. An inborn content-specific mechanism is not needed to account for the fact that humans all have navels; neither is an inborn content-specific mechanism for social reasoning needed to explain why humans typically trade foods or seek what is necessary for warmth. There would undoubtedly be *some* common features even with only generalized learning rules. Presumably this is not what Cosmides and Tooby intend to deny. It must be *some specific* features common across individuals or across cultures that tell in favor of specialized rules for social exchange. Unfortunately, Cosmides and Tooby are not explicit about what these might be. This is an interesting problem, one needing attention. As we return in the chapters that follow to the cases they discuss, we may gain some insight. Still, even for those of us who are skeptical of psychology without specialized rules, we are left with little justification here for the idea of modular and biologically endowed devices.

Cosmides and Tooby offer us a second argument for the same view. It is also negative in its thrust. It is worth quoting at some length:

the Standard Model would have to predict that wherever social exchange is found to exist, it would have to be taught or communicated from the ground up. Because nothing about social exchange is initially present in the psychology of the learner, every structural feature of social exchange must be specified by the social environment as against the infinity of logically alternative branchings that could exist. It is telling that it is just this explicitness that is usually lacking in social life. Few individuals are able to articulate the assumptions that structure their own cultural forms. (Cosmides and Tooby 1992, 208)

Cosmides and Tooby suggest that shared assumptions make possible the communication of specific cultural information. These shared assumptions are what constitute the content behind social exchange. They are the background against which social exchange makes sense. Without these assumptions, there would be no way to learn about our social life. The general line of reasoning should be relatively familiar to those who are aware of the issues concerning the learning of language. It is a line of reasoning with an impressive pedigree, from Immanuel Kant through to Noam Chomsky and beyond. I will not pursue those parallels at the moment, though I will return to them in a later chapter. The problem with this line of thought is reasonably clear. The fact that we have at our disposal cognitive mechanisms that facilitate reasoning about social exchange is something everyone must concede. Were that not so, we certainly would not have international banking systems or international trade. We would not even have markets in which we can buy food or household goods. Even simple bartering would be impossible. We can assume the psychological evidence settles that issue—we need not turn to the *Psychological Bulletin* for definitive evidence. Nevertheless, the fact that we engage in such exchange is not sufficient to show that there must be inherent shared assumptions concerning social exchange that are responsible for the acquisition of those rules. It is possible, so far as that evidence goes, that early learning establishes specialized social rules that we deploy in reasoning about social contracts. Nothing offered in the argument bears directly on *how* the cognitive mechanisms are acquired.

Consider an analogy. We certainly have at our disposal cognitive mechanisms that facilitate reasoning about mathematics. It may be true that there are inherent cognitive schemata that make it possible for us to learn mathematics. Kant certainly thought this was so, and moreover thought it something that could be established a priori. Nonetheless, it is not simply the fact that we can reason about mathematics that, in Kant's view, establishes there are such schemata. Kant attempted to ground the necessity for these prior schemata in peculiar features of mathematical knowledge. Likewise, Chomsky's argument for the importance of an innate grammar depends crucially on both the structure of acquired grammars and on the sorts of inputs on which our learning of language depends. Cosmides and Tooby, by contrast, move seamlessly from the idea that we have some specialized cognitive mechanisms for social reasoning to the conclusion that the acquisition of these cognitive mechanisms depends on the prior existence of foundational social schemata. Somewhat ironically, they do this without much concern for the specific judgments we make. Where Kant thought an argument was needed, Cosmides and Tooby think none is necessary. Where Chomsky thought a proof was needed, Cosmides and Tooby do not. I do not doubt that the conclusion could be true. We certainly do engage in social exchange; perhaps there are specialized and innate cognitive mechanims responsible for this. That does not change the fact that the conclusion is not rooted in any persuasive argument. We have no reason to believe it is true, or even that it is more likely than not.[4] As I'll suggest, this is not at all uncommon for the claims that evolutionary psychology offers to us.

3   Sociobiology and Evolutionary Psychology

Evolutionary psychology recapitulates some of the issues surrounding sociobiology that followed the publication of E. O. Wilson's seminal *Sociobiology: The New Synthesis* (1975). Richard Dawkins evidently characterized evolutionary psychology as "rebranded sociobiology" (see Rose 2000). For those who have read Robert Ardreys's *The Territorial Imperative* (1966), Konrad Lorenz's *On Aggression* (1966), or Desmond Morris's *The Naked Ape* (1967), the Hobbesian vision they share seems clear enough. There is certainly some continuity of vision, but there is not quite an identity of vision (see, e.g., Downes 2001). Evolutionary psychology is not simply sociobiology a quarter century later. Neither are they divorced from one another, intellectually, sociologically, or politically. So, for example, Charles Lumsden and E. O. Wilson (1981, 99) wrote that "The central tenet of human sociobiology is that social behaviors are shaped by natural selection." This tenet of "human sociobiology" is clearly central to evolutionary psychology as well, as it is to some of the most severe critics of sociobiology. It was also embraced by Herbert Spencer, with the assent of Charles Darwin. Still, even though some of the players are the same, and even though there are common views between them, there are also significant differences.[5]

First of all, evolutionary psychologists appeal fundamentally to computational mechanisms. Typically, they draw more from cognitive psychology than from ethology, both in terms of method and theory (see, e.g., Sterelny and Griffiths 1999). Methodologically, evolutionary psychologists approach their problems with cognitive techniques. For example, Cosmides and Tooby focus on performance in reasoning tasks; what is informative is taken to be our ability to solve problems of reasoning. Theoretically, the psychological mechanisms turn out to be the key explanatory factors. The evidence is often of a sort that would make cognitive psychologists happy, or at least engage their professional interest.

Second, and perhaps more important, sociobiology is an enterprise much broader in ambition than is evolutionary psychology. Many of the most striking successes of sociobiology focus on evolutionary models for animal social behavior. There is, naturally, no reason biologists should neglect the explanation of social behavior; and they do not. In the decades following Wilson's *Synthesis*, comparative biology has flourished and benefited from being placed in a more rigorous evolutionary context. Spiders and hyenas turn out to be very interesting animals from a sociobiological standpoint. Human sociobiology may be largely continuous with human evolutionary psychology, but sociobiology has a broader scope.

Evolutionary psychology is geared especially to the human case rather than to animal social behavior. Some advocates of evolutionary psychology do turn to animal behavior, or even animal psychology, but that is typically not their central focus. Whereas some of the most salient successes of sociobiology concern animal behavior—there are fascinating studies of primate social behavior or chimpanzee politics—evolutionary psychology is fundamentally a piece of human sociobiology. Philip Kitcher (1985) derisively called the latter "pop sociobiology." Stephen Jay Gould, with an equally negative tone, called it "pop ethology." Incest was for a time a centerpiece for much of human sociobiology. Here is an example of the way the reasoning went at the time (see Ruse 1982; van den Berghe 1980, 1983): Nearly all cultures treat incest as taboo. So incest taboos are instances of what is called a "cultural universal." It's important to be careful in assessing what this means. Kinship relations are crucial to all societies; kin are restricted in their acceptable sexual relations. These prohibitions are incest taboos. We can, apparently, explain the prevalence of those taboos in terms of evolutionary advantage. Inbreeding has a variety of adverse effects, many of which can be understood genetically. When close genetic relatives interbreed (siblings or first cousins, say), there is a greater likelihood that recessive and deleterious genes will be expressed that would be masked in unions among more distant relatives. In fact, that is something like what we see: mortality rates are markedly higher among children that result from incestuous unions. Given the evolutionary advantages it would offer, the reasoning goes, there must be some evolved mechanism that decreases sexual attraction under circumstances that would be conducive to inbreeding.[6] Since the human family—including both the "nuclear family" and the extended family group—involve close contact, it appears likely that the close contact is the "trigger" that reduces attractiveness. Again, there is some evidence in support of the result, but it offers little, if anything, to the socio-biological speculations concerning the basis of kinship avoidance.

In fact, the appearance of an easy connection between incest and inbreeding is illusory. Questions concerning the "data" are many and probing (see, e.g., Kitcher 1985, 169ff.). Estimates of incest levels vary a great deal, as do the very definitions of incest. Generally, getting reliable estimates of sexual behavior is a difficult matter, and given the social prohibitions on "incest," it would be stunning if empirical tallies did not systematically underestimate the frequency of incest. As with cases of sexual abuse, incest is not something even the victims readily admit. Given the severity of the social sanctions, admitting to incest is something we would not be inclined to do. Sometimes these social prohibitions preclude a wide variety of intimate contact. The severity of the sanction is often stunning, especially for women. Sometimes prohibitions limit the scope to heterosexual intercourse and to specific forms this might take. But incestuous prohibitions are not limited to those genetically related, though that is what would be "predicted." Relations between step siblings are within the scope, typically, as much as relations between full siblings. Sometimes the prohibitions do not reflect the biological divisions, though again this is what would be "predicted." In particular, the proscribed relations often do not apply to those who are biologically kin. Here is Kitcher's conclusion:

> How strong is the human propensity to avoid copulating with those known intimately from childhood? How strong is it in contemporary members of our species? How strong has it been in the past? We simply do not know. . . . Any serious study of incest should recognize the extent of our ignorance rather than rushing to pronounce on the behavioral rule that people must be following. (1985, 274–275)

There are also questions concerning how we could use the data, even if we were convinced of its robustness and reliability. For example, when offspring do not suffer deleterious effects, how many are the products of incest? The simplest question is the most obvious: why are there taboos for something that we naturally avoid?[7] Sociobiology and evolutionary psychology make a good deal out of the pattern we find among contemporary cultures, and of the broader pattern shared among animals. That pattern is complicated to say the least. The most striking problem derives from the clear fact that social kinship is not generally biological kinship. In matrilineal societies, the biological father belongs to a different social unit than his biological offspring; so, although there are limits on sexual and marital relations among clan members, a male may not be prohibited from these relations with his biological offspring—who are, after all, not regarded as his daughters (see Sahlins 1976; Benton 2000). Likewise, in many cultures, females move away from the family unit, and their offspring are no longer considered relatives of those in the original family unit. Once again, the *socially* enforced taboos do not match the *biological* imperatives. We seem to need a different sort of explanation if we are to explain the social prohibitions.

Evolutionary psychology sidetracks many, though not all, of these questions, attempting to focus on what was present in our ancestral groups rather than the confusing mosaic we find in contemporary cultures. This introduces some complications to the questions. If we want to explain some pattern of human behavior, or some human social arrangements, then we need to step back to see what that pattern is, independently of the theory. If we want to know whether selection would favor incest taboos, what is relevant is not the current array of varied social arrangements, but the consequences of incest in the past. If incest taboos extend to adopted children who are not biological relatives at all, that may be because contemporary families do not reflect ancestral conditions. So these arrangements may be little more than side effects of ancestral adaptations that require us to care for kin. Assuming ancestral groups were essentially extended families (which would be likely), then sexual relations within those groups would be expected to have adverse biological consequences. Of course, all this is done in spite of the fact that we do not know the social organization of ancestral groups, and in spite of the fact that extant "primitive" groups do not respect the condition.[8] Whether these taboos now maximize fitness under current social arrangements is not relevant, according to evolutionary psychologists; adaptations need not be adaptive now. What matters is whether the traits in question maximized fitness under ancestral conditions. So, to underscore the point: assume that ancestral groups were organized in a way such that social structure reflected biological relatedness. This might even be true. It doesn't matter that current social structure does not reflect biological relatedness. If ancestral groups were so structured, then selection would have favored an aversion to incest. What is missing is independent evidence concerning the social structure of ancestral groups. This will be a recurring theme in the chapters to follow.

Evolutionary psychology is also, at least superficially, more modest than sociobiology in its pretensions concerning what biology implies for social theory. To be sure, we are the products of our evolutionary history. We are also the products of our social upbringing. Evolutionary psychologists are commonly reticent about claims to genetic determinants of culture, or genetic determinants of psychology. To use a common analogy, assume we represent an organism as a point in space. One dimension of this space is genetic, and the other is environmental. The organism is a product of both its genetics and environment, just as a point in a two-dimensional space must be located in terms of two variables. Knowing your latitude does not suffice to locate you on the surface of the earth. Likewise, knowing only the genetics (or only the environment) will not suffice to locate an organism (see Ridley 1993, 316ff.). The conclusion is supposed to be that evolutionary psychology is not committed to genetic determinism. The analogy is deficient in many ways—mostly, in being two-dimensional—but it aptly illustrates that distinguishing heredity from environment, or genes from their context, is a mistake.[9]

Cosmides and Tooby are content to suggest that psychological theories provide the "foundations" for "theories of culture," and at least in more moderate moments do not claim overtly that they "constitute" alternative theories of culture. In this they appear to be less ambitious than their sociobiological forebears. With this caveat in place, Tooby and Cosmides (1992, 115) continue:

> Nevertheless, increasing knowledge about our evolved psychological architecture places increasing constraints on admissible theories of culture. Although our knowledge is still very rudimentary, it is already clear that future theories of culture will differ significantly in a series of ways from Standard Social Science Model theories. Most fundamentally, if each human embodies an evolved psychological architecture that comes richly equipped with content-imparting mechanisms, then the traditional concept of culture itself must be completely rethought.

The picture turns out to be difficult to articulate cleanly. Certainly, it is not articulated with any precision. As they continue to lay out the project, Tooby and Cosmides say they do not intend to abandon "the classic concept of culture," although they reject the "standard social science model." Instead, they say, they are "attempting to explain what evolved psychological mechanisms cause it to exist" (Tooby and Cosmides 1992, 118). There is evidently some ambiguity, but in at least some expressions, evolutionary psychology does not embrace the aggressive reductionist program of human sociobiology.

4 Sociobiology and Its Critics

E. O. Wilson (1975), in reflecting on the understanding of human nature, mused that "in the free spirit of natural history" humans are, like other organisms, the products of natural history, with their own characteristic suite of adaptations. I don't doubt that this is true. Neither do Wilson's critics. In Wilson's hands, these products of natural history include a variety of features. Humans are aggressive, xenophobic, deceitful and "absurdly easy to indoctrinate." They are also cooperative, caring, and socially connected. In part Wilson pleads that an informed biology should be capable of reforming social science and psychology. This sentiment is certainly shared by contemporary evolutionary psychologists.

Wilson's critics, as is well known (see Kitcher 1985), were vocal and uncompromising. Wilson and his defenders saw them as dogmatic ideologues, pursuing a political agenda at the cost of scientific objectivity. A similar view is often voiced concerning critics of evolutionary psychology (see, e.g., Alcock 2001). In an extended letter to the *New York Review of Books*, the "Sociobiology Study Group of Science for the People" wrote that Wilson's evidence

has little relevance to human behavior, and the supposedly objective, scientific approach in reality conceals political assumptions. Thus we are presented with yet another defense of the status quo as an inevitable consequence of "human nature." (Allen et al. 1975, 261)

The natural and inevitable response to this assault is to counter that the critics have deserted scientific standards and human reason. Of course, that is also the charge offered by the critics of sociobiology. Charles Lumsden and Wilson (1981, 40) deny flatly that "scientific discoveries" should be judged by their political consequences. Wilson underscored the point in *On Human Nature* (1978), though his critics see him as simply repeating the error rather than correcting it. Finally, Wilson's response to his radical critics is to insist that a science of human nature, like any other science, should answer only to the evidence. He says this in protesting the political criticisms:

Human sociobiology should be pursued and its findings weighed as the best means we have of tracing the evolutionary history of the mind. (Wilson 1975b, 50)

Critics again agree on this point, though the friends of sociobiology would insist that the agreement is disingenuous. Here is one response from Stephen Jay Gould (1977b, 258), once a member of the Sociobiology Study Group, and of course a persistent critic of sociobiology:

Scientific truth, as we understand it, must be our primary criterion. We live with several unpleasant biological truths, death being the most undeniable and ineluctable. If genetic determinism is true, we will learn to live with it as well. But I reiterate my statement that no evidence exists to support it, that the crude versions of past centuries have been conclusively disproved, and that its continued popularity is a function of social prejudice among those who benefit most from the status quo.

A similar point is echoed by more recent defenders, such as John Alcock (2001). Having insisted that sociobiologists need not be genetic determinists and need not be adaptationists, he claims that all that is required is "the willingness to test hypotheses about the possible adaptive value of complex social attributes" (Alcock 2001, 217). Critics may doubt whether this describes sociobiology, but it is the view of sociobiologists.

Philip Kitcher points out that the naive appeal to truth and evidence—think of it as the high road—misses the complaint. The fundamental complaint from critics of sociobiology is that the evidence is not there, that the grounds are inadequate, and that the interpretation of the "data" is biased. It is not the willingness to test hypotheses that they see as the problem, but the quality of the tests. Kitcher (1985, 8) puts it, starkly, as a friend of the critics would: "The dispute about human sociobiology is a dispute about evidence." I am unequivocally on Kitcher's side, as it applies to human sociobiology, as well as to evolutionary psychology. Advocates saw sociobiology as insight in the service of truth. Advocates see evolutionary psychology, too, as working in the service of the truth. Critics saw sociobiology as speculation in the service of political ends. As Kitcher (ibid., 9) says, "the truth is rarely pure and never simple." He continues:

when the hypotheses in question bear on human concerns, the exchange cannot be quite so cavalier. If a single scientist, or even the whole community of scientists, comes to adopt an incorrect view of the origins of a distant galaxy, an inadequate model of foraging behavior in ants, or a crazy explanation of the extinction of the dinosaurs, then the mistake will not prove tragic. By contrast, if we are wrong about the bases of human social behavior, if we abandon the goal of a fair distribution of the benefits and burdens of society because we accept faulty hypotheses about ourselves and our evolutionary history, then the consequences of a scientific mistake may be grave indeed. (Ibid.)

The point is correct. Kitcher continues to draw the important conclusions in a compelling way. We do need to attend to the evidence. The evidence unfortunately is often ambiguous. When it is not, we should embrace the conclusion, whatever it is. When the evidence is uncertain, the standards of evidence depend on the risks of embracing the views in question. Risk aversion is not always a failing.

Kitcher (ibid.) explains that when the question is whether to adopt a hypothesis, whether we should do so depends not only on the likelihood of the hypothesis given the available evidence, but also on the costs and benefits associated with embracing the hypothesis if it turns out in fact to be false. Kitcher uses a straightforward analogy. Drug manufacturers insist on higher standards of evidence when there are potentially dangerous consequences from marketing a product. If the side effects from a treatment are minimal, then the risks of promoting a product are minimal, and even a slight chance of benefit makes using the product reasonable. If side effects are dramatic and negative, that changes the equation. We should rationally and reasonably expect better evidence for therapeutic effects in the latter case than in the former.

Kitcher's example is appropriate. It is also very general. Embracing and marketing some new drug treatment is a risky and uncertain enterprise. If the therapeutic effects of a drug are themselves uncertain, and if the potential side effects from a treatment are especially odious, then we should demand better evidence that the therapeutic benefits can be realized before we undertake the treatment. It would be irrational, in fact, not to demand better evidence. The benefits of a successful treatment generally come with a cost that is not merely monetary, and we need to be proportionally certain they can be realized before we bear the cost. Drug companies are not the most palatable example, of course, since their interests are largely monetary.[10] The point applies in the same way to a choice of therapy by a patient, where the costs are personal and the risks are to health and happiness rather than profits. Pursuing any therapeutic course has potential benefits and risks. Humans are, in point of fact, not especially good at weighing the costs and benefits, but it is reasonably clear that certain outcomes are preferable to others. We should look not only to the potential benefits of a treatment, but to the personal costs.[11] A successful treatment might be conducive to survival, but only at a dramatic cost to the quality of life. The cost to the quality of life matters. The point may be even more transparent when the chances of success are remote. If possible but unlikely success comes only at a high cost, we should be reluctant to undertake an aggressive therapy; this means we should require a high level of certainty that it will be successful. So to undertake an aggressive course of chemotherapy, with dramatic suffering, we should expect a greater certainty that the therapy will be successful. If therapy comes at a low cost, then even a long shot is a good bet. So taking aspirin as a prophylactic against heart disease is certainly a reasonable choice, even if the therapeutic effect is small, given that the negative effects are even smaller. Kitcher's point is that a *rational* choice needs to take into account both the costs and benefits, and that as the costs become more unacceptable, it is reasonable to demand more certainty that the therapeutic gains can be reached. If the treatment is likely to impose suffering, then it is in fact irrational not to expect more certainty that the therapeutic benefits will follow. Even this is not all that is required; for whether it is reasonable to undergo some treatment also depends on what would be expected without treatment. There is always a choice, after all. The reasonable choice needs to account for all these factors. To focus only on some of the effects is irrational.

There are some surprisingly simple and general morals to draw from Kitcher's thesis, all widely accepted among decision theorists. First, the decision whether to choose one alternative rather than another depends on the expected benefits as well as the expected costs. A choice with expected costs that are greater than expected benefits is not a rational choice. The decision whether to accept or reject a hypothesis is also a choice. It has attached costs and benefits. Kitcher presents the problem of hypothesis choice as a problem of decision under risk rather than uncertainty, though that does not really affect his central point.[12] Deciding whether to accept or reject a hypothesis (or whether to undergo a course of therapy) is susceptible to treatment under the theory of decision, and the rational decision does not depend *simply* on whether the hypothesis is more likely to be true than false. There are costs and benefits to be weighed here as much as anywhere. So in deciding whether to accept some hypothesis, we should pay attention to the costs and benefits of doing so. To say this is simply to notice that some issues have more human impact than others. Kitcher's point is that when the negative consequences of accepting some conclusion are great, and the conclusion is itself uncertain, then we should demand higher standards of evidence before we embrace it. If we suppose that some hypothesis has a definite chance of being true, given the evidence, then whether we would be rational to embrace it depends proportionally on the consequences of doing so. This includes not only the benefits we might reap if it is true, but the ill effects of making a mistake. Even if we are reasonably certain that the hypothesis is true, if the costs of error would be high, then that counsels caution. The choice of hypotheses is a matter of evidence, but not just a matter of evidence. The consequences of embracing a hypothesis generally matter. If they do not, then they do not. Number theory is like this. Perhaps also some areas of theoretical physics are like this. Typically, though, consequences are significant.

5 Setting the Standard for Evolutionary Psychology

I have so far developed the case for the twin thoughts that although scientific hypotheses *should* rationally be judged only on the basis of the evidence, it is also true that, rationally, we *should* adjust the standards of evidence we require, depending on the impact those views are likely to have. We should expect sound evidence; and, for theories that matter more, we should expect better evidence. We should expect more of theories that matter more. Malthus's theory, for example, was widely regarded as one with considerable support, although the actual empirical credentials of the theory were not especially substantial. To embrace that theory was tantamount to endorsing a set of social reforms that came only at considerable cost in terms of human suffering. On the other side, the failure to embrace well-confirmed theories can have adverse consequences, as did the abject failure of officials in South Africa to accept that HIV was the cause of AIDS. The consequences in both cases were not good. It was another half-century before the Malthusian mistake was even partially rectified. We still live with some of the consequences. It may take longer to rectify the mistakes of the South African government.

Box 1.1
Decisions under risk and uncertainty

Conceived as a problem of decision making under risk, this sort of problem is often represented as a decision with four components. We can assume for simplicity that the choice is simply whether to accept or decline a given course of therapy, and that the patient either recovers or does not. The expected benefit from a proposed course of therapy depends on two factors. We suppose there is some definite probability of recovery under a given therapy. There is also a probability of not recovering under that therapy. What is called the "expected utility" of the choice is then the sum of the values of the outcomes weighted by the respective probabilities. The outcomes are complex; they involve both costs and benefits. But it is not difficult to draw some qualitative morals. If the suffering under the treatment is considerable, then it is a near approximation to conclude that a proposed therapy would be expected to be desirable provided the likelihood of recovery is greater than the ratio of the cost of therapy to the benefit of therapy. As the negative side effects—that is, the costs—become more pronounced, the utility of a proposed therapy will be positive only if the likelihood that the therapy is effective is proportionally greater.

Things are even more complicated than this simple point allows. Whether it is reasonable to undertake a proposed course of therapy does not depend simply on the expected utility of the therapy; it also depends on the expected utility of declining a therapeutic alternative. That is, whether it is reasonable to undertake a therapy depends on the expected utility of the therapy, but also on the expected utility of declining the therapy. There is always some chance of recovery without therapy and a likelihood of some suffering even with the therapy. These can be used to define the expected utility of declining a course of therapy. What we get from decision theory is even more cautious than Kitcher suggests. We need a comparative judgment. The issue is one of choice, and which choice is better. As a result, we need to factor in not only the chances of recovery under therapy, and the costs of therapy, but also the chances of recovery without therapy, and the benefits of declining some treatment.

	Recovery	No recovery
Adopt a course of therapy t	Probability p of recovery × (benefit of recovery − cost of suffering under t)	Probability $1-p$ of non-recovery × (benefits w/o recovery − cost of suffering under t)
Decline a course of therapy t	Probability q of recovery × (benefit of recovery − cost of suffering w/o t)	Probability $1-q$ of non-recovery × (benefits w/o recovery − cost of suffering w/o t)

It is reasonable and rational to be responsive to the evidence relevant to a scientific view, both positive and negative. To think otherwise would be an affront to science and rationality. It is reasonable and rational to be responsive to the social impact of scientific views. To think otherwise would require that we reasonably or rationally deny that there is a social impact, and that too would be an affront to science and rationality. Malthus wrote with no illusions about the consequences of his views. He knew there would be suffering, but he thought that suffering would be ultimately exaggerated by the continuation of the poor laws, even in an amended form. He thought suffering would be ameliorated if we abandoned support for the poor. That is not true. Those who denied that HIV was the cause of AIDS acted in good conscience, thinking that the alternative enhanced rather than alleviated suffering. That also is not true. We should not pretend that science exists apart from its social applications. That view is contradicted by not only the manifest social impact of early genetics, but also of nuclear physics. These sciences did, after all, have as their respective offspring eugenics and the threat of nuclear war. To deny that science often has profound social consequences would be to ignore the social benefits we have collectively garnered, and might yet gain, from genetic therapies and nuclear power plants. Even the simple Darwinian recognition that humans are descended from other animals, that we are but modified apes, has marked consequences. Without that, much of modern medicine would make little sense. We would have, for example, no reason to think that drugs tested on animals would work on humans as well if we did not know that we share certain physiological mechanisms. We would be irrational to ignore the applications of a scientific theory just as we would be irrational to ignore the evidence for or against it.

Evolutionary psychology does have as a core part of its agenda the overthrow of much of contemporary social science and psychology, including the model on which these sciences supposedly depend. That would be an important consequence, if it happens. It could also have important social implications beyond the academy. Doubtless, evolutionary psychology would not have the dramatic social consequences that Malthusianism promised. It is nonetheless far from uncontroversial in its implications. To take one of the most contentious cases, Randy Thornhill and Craig Palmer's *A Natural History of Rape: Biological Bases of Sexual Coercion* (2000) argues that rape is a behavioral strategy that enhances male fitness. To make this argument they assume that females should be selective in choosing mates, and that males should not. Males would then be inclined, naturally, to engage in rape whereas females would not. They also say, a bit paradoxically, that males should prefer sex with fertile young women, which of course entails that men would be selective rather than indiscriminate, for which I see no evidence at all. The evidence for the advantages of rape was largely drawn from examples of "rape" among nonhuman animals. The appeal to mallard ducks and scorpion flies is of dubious relevance to human rape, which, though a sexual act, is also a crime of violence.[13] Thornhill and Palmer were careful *not* to take their analysis as any kind of moral mandate for rape; it was, rather, an evolutionary explanation of why men rape. They insist that their interest is in helping rather than harming women. Nonetheless some critics understood their analysis to justify rape, and, at least, to give some comfort to rapists. Thornhill and Palmer's work may be an extreme case, where the implications are dramatic and the politics especially volatile. I have already observed, though, that evolutionary psychology is not lacking in ambition. What the case of Thornhill and Palmer does illustrate is that the social implications of evolutionary psychology are not lacking. That counsels caution. David Hume noticed the relevance of a theory's social implications to its being advocated. In his *Enquiry Concerning the Principles of Morals*, he wrote:

And though the philosophical truth of any proposition by no means depends on its tendency to promote the interests of society; yet a man has but a bad grace, who delivers a theory, however true, which, he must confess, leads to a practice dangerous and pernicious. Why rake into those corners of nature which spread a nuisance all around? Why dig up the pestilence from the pit, in which it is buried? The ingenuity of your researches may be admired; but your systems will be detested. (Hume 1751, §9, part 2)

So we might reasonably and prudently expect correspondingly sound evidence, insofar as we think the implications of evolutionary psychology are of social importance. Mallards and scorpion flies are too remote to be good evidence concerning rape in human societies. From an evolutionary perspective, these are remote analogies, when what we require is more direct evidence. It is at least "bad grace" to promote such analogies in the absence of evidence.

On the other side, we should not demand the impossible. Michael Ruse (1979a, 21), a more or less conservative voice, says this: "as sociobiology is part of the evolutionary family, we ought not to judge it by standards more strict than we would apply to the rest of evolutionary theory." I assume that evolutionary psychology would get a similarly favorable treatment from Ruse. This much is at least clear. It is unreasonable to demand what it is otherwise unreasonable to expect is possible. It would be unreasonable to set standards of evidence that no one could meet. The goal is after all to reach a reasonable assessment, not to peddle some fatuous form of philosophical skepticism. Such skepticism would be no less a philosophical nuisance, even if it might be a relatively harmless one. If we set the standards excessively high, we can surely rule out evolutionary psychology. But we might thereby also rule out all of evolutionary biology. That would make the standards suspect. The threat is in setting the requirements on an acceptable explanation so high that they would rule out work in evolutionary biology and ethology. Evolutionary psychology would thus be ruled out by excessively high standards of evidence; but so would the background in which the discussion takes place. That is the key point. It is evolutionary biology that defines the context in which the adaptive claims of evolutionary psychology should be assessed. The standards we should use are evolutionary standards. Distorting that context would correspondingly distort the issues. That is a consequence we should carefully avoid.

In what follows, I will press that we should reject the pretensions of *evolutionary* psychology largely as unconstrained speculation, as claims ungrounded in evolutionary history. The moral I offer is certainly a skeptical one. I will argue that we have no credible reason to embrace the explanations offered within evolutionary psychology. I advance no alternative explanations; I do not generally claim that these evolutionary hypotheses concerning human psychology are *false*. I do think that some do not warrant serious consideration, for lack of evidence. For all intents and purposes, we can dismiss them. Some may eventually yield to empirical discoveries. If this is so, then so be it. I hope, in part, to lay out the kind of evidential standards that should be met. These questions are my central focus. I argue, fundamentally, that the empirical and historical record gives us no reason to accept the prevailing views in evolutionary psychology.

I suggest, but do not demonstrate, that in the end we are unlikely ever to have the sort of evidence that would be required to make it reasonable to embrace the hypotheses of evolutionary psychology. Mine is a skepticism that also embraces the alternatives. I doubt we will be in a position to know that some nonadaptionist alternative is true, in the human case. It is important that I do not intend to plead for higher standards than we would expect of evolutionary explanations generally. I will insist, if nothing else, that the standards I embrace are not unreasonable. The way to this conclusion is simple: I will draw the standards from respectable work in evolutionary biology. Ruse's standard will be the working standard here, even though the results are not his results. Evolutionary hypotheses are subject to a variety of empirical tests, though rarely the kind of test that would warm the heart of more narrow experimentalists. The heart of the book is organized around three different approaches toward empirically evaluating evolutionary explanations. These include what is called "reverse engineering," the inference from function to cause (chap. 2); another alternative is to infer effect from the relevant causes, an approach that reflects the dynamic perspective of much of evolutionary population biology (chap. 3); finally, I will turn to analyses designed to disentangle history from structure, which depends on disentangling historical ancestry (chap. 4). I believe that the later analyses are the most powerful. They are more likely, as analyses of adaptive function, to discriminate between adaptation and other causes. The three are commonly appealed to within evolutionary biology, and their limitations are broadly understood within evolutionary biology. I am convinced that my applications lie within the boundaries that are and would be accepted by evolutionary biology.

Generally, the task involved in assessing an evolutionary explanation— whether of human psychology, clutch size in birds, or flower structure in orchids—is one of assessing the historical antecedents and context, given both the contemporary forms and the relevant historical record. These antecedents in turn provide evidence essential for a defensible evolutionary explanation. Evolutionary psychology offers proposals *within* the broad framework of evolutionary theory. Evolutionary psychology is not a challenge to evolutionary theory; it is a challenge to contemporary cognitive psychology. I will work within the framework of contemporary evolutionary theory. Of course, if I misrepresent those standards, that is a complaint that would be decisive against everything that follows. The assumption I work with is that the sorts of conjectures defended by evolutionary psychologists should at least be held to the same standards that are properly demanded of evolutionary explanations. If we are offered evidence for an hypothesis that would not suffice as evidence for an evolutionary explanation of clutch size in birds or flower structure in orchids, then we should reject the explanation. Given these standards, I think the empirical and historical record is insufficient to support the explanations offered within evolutionary psychology. If we were given an explanation for clutch size or flower structure based on similar evidence and reasoning, we would and should reject it as premature and inadequately supported. It would, in fact, not be the sort of evidence we would expect for a hypothesis that is a serious contender. We should be no less uncompromising in our expectations for evolutionary psychology. For those who maintain that a higher standard of evidence would be appropriate, in light of the implications evolutionary psychology has for human well-being, I would ask them to remember than my verdict will finally be negative. With a lower threshold, I will press that evolutionary psychology fails. To raise the bar would only strengthen my case, but if I am right, it is not necessary. Of course, we might also choose to lower the standard. To do that would be, as I've argued, an affront to science and rationality. As Hume saw, it is also bad taste. If it is "bad grace" to defend a theory that leads to a "practice dangerous and pernicious," even when the theory is justified, it is worse grace still to advance such a theory when the evidence is lacking.

2   Reverse Engineering and Adaptation

1   The Appeal of Reverse Engineering

The study of adaptation within an evolutionary framework inevitably involves inferring historical processes from contemporary products. The aim is to understand the historical sequence and causal antecedents, revealing prior conditions as determinants of contemporary patterns. In short, we infer cause from effect. This is a difficult task, but certainly not one that is impossible. Direct evidence concerning ancestral environments, variation, social structure, and other relevant features are often not available, though they sometimes are. Some biologists, especially evolutionary ecologists and behavioral ecologists, focus first on questions of current form and function, abstracting initially from the historical paths that produced what we see. Like archaeologists, these evolutionary biologists begin with the artifacts and use them to unravel the history, though in this case the designer is Mother Nature, and the product is natural design. Models developed by biologists for parental investment, for reciprocal altruism, and for inclusive fitness are all elaborations or extensions of the theory of natural selection, and all are various species of optimality models. They assume that natural products are well suited for their functions. In such applications, and also in cases of reverse engineering, the question at the focus of investigation is not initially historical, but rather concerns the extent to which a trait optimizes fitness among a specified set of variants and within a specific environment. The standard for fitness in all these cases becomes optimality of design, and in reverse engineering, this is gauged by current utility.[1]

Reverse engineering thus is a matter of inferring adaptive function from structure. This leaves it deliberately ambiguous whether *function* is intended as *historical function* or as *anatomical function*. In the first case, the function of a trait is what it *was* selected for in the past, however it may feature in the current economy of the organism. In the latter case, the function involves how the trait *is* currently used, whatever its historical function might have been. In its most controversial forms, reverse engineering aims to infer the historical causes from observed organic form. That is, the goal is to infer the historical function, and then to use this in order to explain current form. If *adaptive thinking* begins with the ecological "problems" an organism confronts and explains or infers the likely "solutions" based on the problems, reverse engineering turns the reasoning around, beginning with the "solution" and inferring what the ecological "problem" must have been if those traits were to evolve (see Griffiths 1996).[2]

It is important to distinguish, sharply, the direction of inference. David Buller (2005) distinguished *reverse* engineering from *forward* engineering—the latter being what I've called "adaptive thinking." Forward engineering involves inferring a mechanism from the demands imposed by the environment. Reverse engineering involves inferring the task from the design (see Buller 2005, 92–93). Forward engineering is the task faced by an architect building a bridge, given a river in place, with various constraints imposed by the geological structures. Reverse engineering would be inferring the river width and constraints from the blueprints. Buller is less than clear in the way he describes the differences, but I think the distinction he wants is this one. He says at one point, in criticizing Gould:

Adaptive reasoning in Evolutionary Psychology must be understood in this context of anticipated discovery. For the psychological adaptations are *inferred from* adaptive hypotheses. (Buller 2005, 90)

If this is *functional analysis*, or adaptive thinking, that is one issue. I'll deal with it in the next chapter. Immediately following that, Buller (ibid., 92) identifies functional analysis with reverse engineering. Insofar as we are interested in what is needed to justify evolutionary and adaptationist claims, these differences matter. They concern what counts as evidence, since the claims are different. So let's sharpen the edge for the moment and focus on reverse engineering.

John Beatty (1980, 533) offers an elegant example of reverse engineering. This is a species of optimality reasoning, whose goal, he says, is to explain the "morphological and behavioral design of organisms." Beatty draws on a case describing the predatory behavior of praying mantids. Mantids have a forelimb with a claw, enabling them to grasp prey. Beatty explains that Holling (1964) examined the maximum size prey that could be efficiently grasped and retained by one species of mantids. Assuming that larger prey are a more efficient food source and that efficiency in feeding will maximize fitness, Holling inferred that this should be the favored size for a food source. Notice that we begin with the anatomy, with biological form, and infer the adaptive function.

We do *not* begin with an adaptive problem—say a menu of available prey of various sizes, perhaps with different nutritional levels—and infer what would be an optimal form for the claw in order to utilize the prey that are available. This latter approach would be adaptive thinking. Instead, we look first to the mantid's claw structure and infer what its function must be or must have been. One virtue of these sorts of models, and optimality models generally, is that they allow us to sidestep the relative paucity of evidence we sometimes find when we want to understand historical processes and try to infer something about the process from the outcomes. When it turns out that the mantids do preferentially attack prey of the predicted size, Beatty (1980, 545) appears willing to infer that the structure and the behavior are the consequence of natural selection. Given the fit between form and function, we infer that the form is a consequence of natural selection favoring that form. We use current form to infer the historical function. This is reverse engineering in a historical application.

The general strategy of the reasoning should be familiar, even to those who are not advocates of evolution. It was an integral part of natural theology in the late eighteenth and early nineteenth centuries to argue for the existence of a designer from what was called the "evidence of design." The archbishop William Paley famously argued that current evidence of complex design requires us to accept that there is a designer who is or was the cause of the observed structures. His most famous example is that of the eye. He contended that the structure of the eye is so intricately fit to its function that we are compelled to infer there was some intelligent cause. I do not want to pursue Paley's

Figure 2.1
A praying mantis. There are over 2,000 species of mantids, varying considerably in size. From http://www.mass.gov/agr/.

argument here. I am content to notice the logical form. We begin with the effect. In Paley's case this is the fit of the eye to its function. He then explains this fit in terms of the ultimate cause that shaped its function. This too is reverse engineering, though in a strikingly different setting and with a different intellectual agenda.

In *Darwin's Dangerous Idea*, D. C. Dennett (1995, 233) applauds reverse engineering as "the crucial lever in all attempts to reconstruct the biological past" and claims this is *the* central "feature of the Darwinian Revolution: the marriage, after Darwin, of biology and engineering" (ibid., 186).[3] Here's one example he uses:

Did *Archaeopteryx*, the extinct birdlike creature that some have called a winged dinosaur, ever really get off the ground? Nothing could be more ephemeral, less likely to leave a fossil trace, than a flight through the air, but if you do an engineering analysis of its claws, they turn out to be excellent adaptations for *perching on branches*, not for *running*. An analysis of the claw curvature, supplemented by aerodynamic analysis of archaeopteryx wing structure, makes it quite plain that the creature was *well designed for flight*. (Ibid., 233)

As a merely historical matter, reverse engineering is surely not the central feature of the Darwinian revolution. Darwin hardly emphasized reverse engineering, though he held a central place for adaptation. This is an issue we can defer. More important, I think it is less than clear that *Archaeopteryx* was a full-blown flyer, and we will shortly return to that case in much more detail. Reverse engineering may, in any case, be an important approach to the understanding of adaptation, even if not central to the Darwinian argument. I am interested for the moment in the inferential structure of reverse engineering, and that is displayed clearly in Dennett's reasoning, as much as in Paley's. Reverse engineering is designed to take us *from* structure *to* function and then *from* function *to* history. We are supposed to infer that mantid claw structure is designed for specific prey sizes, and therefore that it is an adaptation for capturing selected prey. Likewise, we are supposed to infer that *Archaeopteryx* was airborne from the structure of its foot, and therefore that *Archaeopteryx* was adapted for flight.

Tooby and Cosmides, similarly, claim that the evolutionary process is analogous to the intentional construction of devices, and so can be understood in terms of reverse engineering. This is, of course, the heart of Dennett's version of Darwin. In the broadest terms, though, the problems "solved" by evolutionary design are limited by the fact that no design is successful unless it either provides a solution to a problem that enhances an individual's fitness, or offers a solution to a problem that increases the fitness of kin. Here is what Tooby and Cosmides (1992, 55) say:

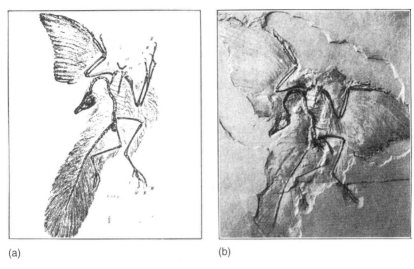

(a)   (b)

Figure 2.2
Archaeopteryx lithographica. The Berlin Specimen. On the left, an engraving following Steinmann-Doderlein, ca. 1884. (The right wing is missing.) On the right, a photograph of the specimen. From http://www.tccsa.tc/images/archaeopteryx.jpg/.

At its core, the discovery of the design of human psychology and physiology is a problem in reverse engineering: We have working exemplars of the design in front of us, but we need to organize our sea of observations about these exemplars into a map of the causal structure that accounts for the behavior of the system.

Applied as a methodological tool for evolutionary biology, then, we can begin with a behavioral pattern, and infer "the causal structure" responsible for the behavior we see; for an evolutionary psychologist, seeing how we behave informs us what the evolutionary or historical function was of the mechanisms governing our behavior. The recognition of the historical function helps us "to organize our sea of observations." In the case of evolutionary psychology, all that changes is that *we* are the organism of interest and that *our own* behavior is what we seek to explain.

Here is another example, of a quite different sort, one that is commonly cited by evolutionary psychologists. M. Profet (1992) suggests that "pregnancy sickness" is an adaptation designed to limit maternal ingestion of substances that might harm a developing fetus (see also Alcock 2001, 195ff.). The key thought is that plants contain an array of toxins to repel herbivores; in some cases, the function is to regulate when fruits are eaten, and in others the function is to discourage herbivores altogether. So leaves are typically bitter, which should repel herbivores. Similarly, unripe fruits are less desirable, whereas ripe fruits are advertised by color and enhanced with sugars. Fruits are vehicles for the dispersion of seeds, from the plant's point of view. Leaves are the means for producing energy, again from the plant's point of view. Plants are thus led to "design" their fruits for consumption and to "design" their leaves to deter herbivores. Yet again, since seed predators are a problem, plants invest in toxins to deter them. The Kentucky coffee tree, *Gymnocladus dioicus*, is laced with toxins to prevent ingestion by predators. Caffeine and nicotine are toxins, which is why eating raw coffee seeds is not a good idea and why smoking cigarettes is an equally bad idea. Apple seeds contain arsenic to deter seed predators, though the seeds are also wrapped in a fruit to encourage animals that will disperse the seeds without consuming them. Profet (1992) supposes that humans must have evolved in response to plant toxins, though we often consume them. Assuming these toxins are more of a problem for a developing fetus, he concludes that it would be natural for a pregnant female to develop an aversion to the toxins in order to protect the fetus. Since "pregnancy sickness" does limit what women can eat, all we need assume is that our bodies are inclined to reject what is unhealthy. In this case, what is unhealthy is what would be harmful to a developing fetus. Plants toxins do dissuade animals from eating them, and so the aversion to eating these toxins might appear to be protection.[4] So spicy foods, which are often mild toxins, are not favorites early in pregnancy. This, at least, is the theory.

Tooby and Cosmides often recount Profet's explanation of "pregnancy sickness," as do others. Once again, here is something Tooby and Cosmides (1992, 77) conclude about reverse engineering:

This form-to-function approach is just as productive as the others because it leads to the prediction and organization of previously unknown facts, usually about additional design features of the organism as well as about the recurrent structure of the world.

The well-known objection to this approach, due to Richard Lewontin and Stephen Jay Gould, is that this is little more than the telling of "just so stories." I'll return to that complaint below in section 3. Tooby and Cosmides suggest that, although it is to just such cases that Lewontin and Gould's attack might apply, it would apply only if the facts about the environment, organism, and the target feature "were known in advance," and this is not so. Why they say this is less than obvious.

2   The Case of *Archaeopteryx*

T. H. Huxley was Darwin's most forceful and colorful advocate in nineteenth-century England, self-styled as "Darwin's Bulldog." In the late 1860s, Huxley brilliantly argued, on the basis of the morphology of birds, that the ancestor of birds must have been a dinosaur. The discovery of *Archaeopteryx* in 1861 provided the crucial intermediate form.[5] There are now several specimens known, dating from the upper Jurassic (c. 145 million years ago). *Archaeopteryx* is a stunning example of an intermediate form, though generally classed as a bird. It does share a variety of features with birds. Most famously, it has feathers and wings. Its other characters are more reptilian than avian. It has teeth, enlarged eyes, a long tail, a brain structure that is more reptilian than avian, and a variety of skeletal characters common with its reptilian ancestors and lacking in birds.

There are two broadly different theories, classically considered, for the origin of bird flight and the ancestry of *Archaeopteryx*. Both date from the nineteenth century. One goes from the ground up. The other goes from the trees down. Though this is a common and useful way of depicting the approaches, we should not let it obscure the primary issue, namely, how the flight stroke, one generating both lift and thrust, evolved. Still, the alternatives do offer broadly different approaches. The place of *Archaeopteryx* in the history of avian flight is connected to the alternative we choose.

The first, "cursorial," theory begins with agile and fast terrestrial dinosaurs, moving very quickly on two legs. As predators, these dinosaurs are creatures of pursuit rather than stealth. Primitive proto-wings, on the original cursorial view, would supposedly have served to enhance pursuit. The most glaring problem with the theory in its original form is that pursuit likely would be not enhanced but impeded by using protowings. Very fast bipedal predators do not

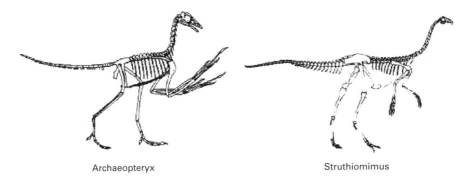

Archaeopteryx   Struthiomimus

Figure 2.3
Archaeopteryx is actually more similar to reptiles than to birds, though it shares some characters diagnostic of birds. The figure on the left depicts *Archaeopteryx*, and that on the right a dinosaur. Ostrom (1976) found ten features that just the skull of *Archaeopteryx* shares with theropod dinosaurs. There are numerous features it does not share with modern birds, as is clear from a comparison of these organisms. (Modified from Ruben 1991, who drew material for the figure from Ostrom 1976 and Russell 1972.)

use wings for propulsion. Some birds that are cursorial predators, such as road-runners in the western United States, even have dramatically reduced wings; certainly, this reduction does not come at the cost of efficient pursuit. J. H. Ostrom (1974) has since revived and revised the cursorial theory, though with a twist that treats wings as predatory aids rather than as propulsion devices. On Ostrom's account, they serve as nets for small prey. More recently, K. Padian (2001) also treats the precursors of birds as active terrestrial forms. The view, we'll see, has considerable attraction.

The second, "arboreal," theory, begins with tree-dwellers, leaping from tree to tree. The development of wings elongates the jump and reduces the rate of descent. So on this view, the ancestors of birds would glide, like flying squirrels or frogs. As gliding improved, a flight stroke would allow for propulsion and not just gliding. Walter Bock (1965, 1986), for one, once suggested something like the arboreal theory is true. On the view he describes, the avian ancestors went through a series of stages: beginning as bipedal, they became arboreal; once arboreal, they went from leaping to parachuting to gliding; in the end, the result was sustained and active flight. The crucial progression is from the trees down.

Archaeopteryx on either view is an intermediate form. On the arboreal theory, *Archaeopteryx* is an active flier, well past the gliding stages. The wing structure of *Archaeopteryx* is certainly consistent with this view. On a cursorial theory, *Archaeopteryx* is an active terrestrial predator, not yet an active flier. Earlier I took note of Dennett's claim that *Archaeopteryx* was adapted for flying, an arboreal form. The capacity of *Archaeopteryx* for active flight has been a matter of persistent controversy. Dennett's is an arboreal model. Once again, here is what Dennett offers: "if you do an engineering analysis of its claws, they turn out to be excellent adaptations for *perching on branches*, not for *running*." It's not clear what exactly he has in mind as an "engineering analysis of its claws" or why this favors an arboreal hypothesis, but it is clear what he thinks follows. The evidence, for the most part, points the other way. A broad consensus that the cursorial model has more support emerged at the 1984 International *Archaeopteryx* Conference in Eichstätt (see Hecht et al. 1985; however, see Ruben 1991 for a dissenting view). *Archaeopteryx* is certainly bipedal. It has a robust pelvic girdle and limbs suggestive of a terrestrial lifestyle (see figure 2.3). Although it evidently was capable of powered flight, its capabilities were likely limited. It lacks, among other things, a sternal keel, which would provide an anchor for the muscles needed for flight. Lacking a keel would compromise the power of the flight stroke. The wing structure, moreover, apparently does not offer enough strength to sustain flight; and the tail is not right. These are certainly design considerations, so they actually fit with Dennett's overall view of the importance of reverse engineering, even if they do not fit with his own view of *Archaeopteryx*. They do suggest that *Archaeopteryx* was at best an awkward flier.

Dennett, as I've said, appeals explicitly to the claws. They are properly called "unguals" in birds, though "claw" is accurate enough and more familiar. Figure 2.4, adapted from Ostrom (1974), is illuminating. In it, Ostrom compares unguals from a variety of birds. Ostrom points out that, in terms of shape, the unguals from *Archaeopteryx* actually resemble ground-dwelling birds more than they do those from perching birds or predatory birds. The curvature is relatively flat, and the small nodes at the base of the structures are less developed in *Archaeopteryx* than in either perching birds or predatory birds. Those structures—reduced or lacking in *Archaeopteryx*—actually facilitate perching. The overall structure of the foot is probably more important; perhaps that is what Dennett meant to appeal to, since it traditionally played a prominent role in pressing that *Archaeopteryx* was arboreal. One of the key features is the reversed large toe, or "hallux." Like many modern birds, *Archaeopteryx* has one toe oriented to the rear and three to the front. In passarines, or perching birds—the familiar finches, pigeons, and parrots, for example—this allows for a firm grasp, as the hallux is fully opposable. In *Archaeopteryx*, too, it looks as if it *must* be an adaptation for grasping branches, just as it is in modern birds. Perhaps, as I've said, these are the sorts of design considerations that entice Dennett. In the case of *Archaeopteryx*, the appearance is deceptive.

Here's first the moral I would draw. Engineering design can certainly be an important feature in evolutionary discussion, but it is not as simple as Dennett suggests. The considerations of design are complicated. It would certainly not be irrelevant if the unguals of *Archaeopteryx* were well adapted for perching; it is also not irrelevant that they are actually similar to those in terrestrial birds. And, it is hard to find either of these facts wholly compelling in light of other design considerations. The *Archaeopteryx* wing is not as well adapted for flying as are the wings of pigeons, either; still, it did provide lift even with a weaker flight stroke. (But then pigeons are extraordinarily good fliers, widely represented, and so that may be an unfair comparison.) Again, we might not find this compelling. *Archaeopteryx* is, after all, to serve as a transitional form. So if its wings are not as fully perfected for flight, its feet might not be as fully perfected for perching either. In that case, of course, the inference from reverse engineering would be weaker rather than stronger. On the basis of reverse engineering alone, the case concerning *Archaeopteryx* is hardly compelling, even if it is in parts suggestive. The structure seems simply to underdetermine function.

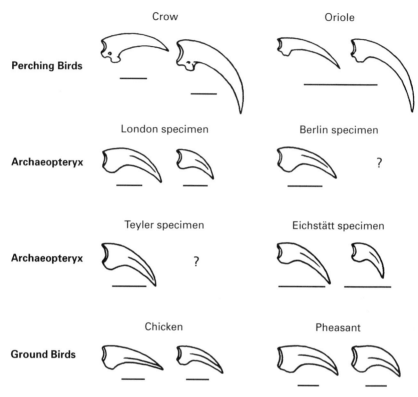

Figure 2.4

A comparison of ungual structure in birds and *Archaeopteryx*. The key feature is the resemblance of the unguals in *Archaeopteryx* to the ground dwellers rather than either perching or predatory birds. The nodes at the base of the unguals for the predatory and perching birds facilitate perching, and are absent in *Archaeopteryx* as well as the ground birds. So ungual structure actually supports a nonarboreal lifestyle for *Archaeopteryx*. (Figure modified slightly from Ostrom 1974.)

As we broaden our perspective, we can get greater traction with the problem. Return briefly to the structure of the foot. *Archaeopteryx* certainly has a foot suitable for grasping, with its opposable "thumb," or hallux. The foot of *Archaeopteryx* might well have been used for perching. Doubtless it was, in fact. But that does not show that it evolved for perching. Chickens perch too, after all, though for the most part they are ground birds. We use our thumbs for grasping pens, but that was certainly not what they evolved for. The crucial facts lie elsewhere, beyond considerations of design. It turns out that the overall foot structure is essentially the same as the pattern found in theropods, the cousins of *Archaeopteryx*. In *Ornitholestes* and *Velociraptor*, the foot structure is similar and the function is clearly to grasp prey. However, the foot will not facilitate perching. *Ornitholestes* and *Velociraptor* were predators. *Archaeopteryx* too has teeth, which supports a predatory lifestyle. As Ostrom (1974) points out, even the most dedicated arborealist would not imagine that the theropods *Tyrannosaurus* or *Allosaurus* habitually perched in trees, though they too have the reversed hallux. Here's why looking at theropods matters: Dennett's thought that the foot of *Archaeopteryx* is built for perching is undercut by the fact that (other) theropods have a similar foot structure, and that the structure evolved apart from, and prior to, any arboreal life style. Even if the foot were well adapted for perching, that is not what it evolved for. That is not its evolutionary function, whatever it might be used for.

That would be decisive if birds—and crucially *Archaeopteryx*—are modified theropods. This seems likely, although the matter again is hardly settled. There are two broad proposals in play relevant to the phylogeny of birds. One makes birds descendants of therapod dinosaurs in the later Cretaceous (c. 100 mya). The other makes birds descendants of archosaurian reptiles in the late Triassic (c. 200 mya).[6] On the latter view, you can think of birds as cousins of dinosaurs rather than offspring. On the former view, birds are, metaphorically, the offspring of dinosaurs. The latter view is not lacking empirical motivation. The most birdlike dinosaurs postdate *Archaeopteryx* by roughly fifty million years; this is based on pelvic structure, which may be misleading. In any case, *they* are not the ancestors of *Archaeopteryx*. Perhaps more fundamentally, the digits in the theropod hand are 2, 3, 4 whereas in the avian hand they are 1, 2, 3. This means that the digits lost are different in the two lineages, which counts against any simple line of descent. Reasons like this lead no less an authority than Ernst Mayr to be inclined to favor the view that birds are more ancient, and, not coincidentally, that *Archaeopteryx* is capable of flight (see Mayr 2001, 144; see also Lande 1978 on limb loss in tetrapods).

The view that birds are modified theropods is nonetheless the more common view, especially among cladists. I expect it is the correct view. The pelvis and legs of birds and thecodont dinosaurs are very similar. In recent years, we have seen a number of dramatic discoveries in China that add a decisive piece of evidence. The fact that *Archaeopteryx* has feathers is often taken as sufficient reason to include it among the birds, but whether we choose to do that or not, the feathers in *Archaeopteryx* have deep similarities to those in modern birds (see Prum and Brush 2002). Feathers are very complex structures. For such similar structures to evolve in diverse lineages would be unlikely, though of course not impossible. The recent paleontological discoveries show that other theropods from the early Cretaceous have structures that are likely "primitive" feathers (see Chen, Dong, and Zhen 1998; Xu, Tang, and Wang 1999a, 1999b; Padian 2001; and Xu, Zhou, and Prum 2001). There are now at least eight such forms that have been described, one suitably called *Protoarchaeopteryx*. *Protoarchaeopteryx* is unambiguously *not* a bird. It seems certain, now, that the origin of feathers lies in nonarboreal theropod dinosaurs. Of course, if feathers originated in terrestrial theropods, then both *Archaeopteryx* and modern birds must be theropod descendants. The key thought is that generality suggests priority: if feathers are common in a broader taxonomic category (for example, theropods as well as birds), then feathers must have evolved prior to other features, such as active flight, that are limited to a single group within the clade (for example, birds). This is particularly clear when the broader characteristic is clearly adaptive for the feature specific to the more limited group, as is certainly true here since feathers facilitate flight even if they originally evolved for some other purpose. These are issues we will return to, especially in chapter 4, when the relevance of phylogenies to evolutionary history is in focus.

Feathers thus provide evidence that the foot of *Archaeopteryx* is not designed for perching. This might seem paradoxical; but it is true nevertheless. The presence of feathers in earlier theropods tells us that birds are descended from theropods, and that *Archaeopteryx* is as well. These theropod ancestors were terrestrial and also have the foot structure of *Archaeopteryx*. Even if *Archaeopteryx* chose to sit in trees, as it might have, the foot is still a derived structure and an adaptation for a terrestrial rather than an arboreal form of life. *Archaeopteryx* is thus a modified theropod and also a proto-bird. As a modified theropod, it is likely a terrestrial predator, for whom flight was incidental. Of course, flight is hardly incidental for birds, but that is the key point. *Archaeopteryx* is not a bird; it is not even a poorly designed bird.

I hope that the point of this diversion into the evolution of *Archaeopteryx* is now coming into focus. The key point Dennett wanted to make using the case of *Archaeopteryx* was that we could use information about something's current structure to infer its function, and from there infer something about evolutionary history. The structural information is ambiguous. The inference is tenuous at best. At worst it is simply an error. But the deeper point is one about the reconstruction of evolutionary history and the role of structure. It is not that structure is irrelevant, or that design considerations are irrelevant to evolutionary history. They are not. The point is that unless supplemented and augmented in a variety of ways, considerations of design are inconclusive and likely misleading. This leads to a second moral, also suggested by the case of *Archaeopteryx*. We should not repudiate analyses of design, but seek supplementation from other more specifically historical approaches. Knowing the mechanisms can tell us what an organism can do; the history can tell us whether the function is sufficient or necessary to explain the structures we see.

This is what Paul Griffiths (1996, 515) calls the "historical turn," the idea that "adaptive generalizations of this sort cannot explain form except in conjunction with a rich set of historical initial conditions." The historical turn is crucial for evolutionary reasoning. Form follows function, but only contingent on history.

3   The "Spandrels" and Adaptationism

Stephen Jay Gould and Richard Lewontin (1979, 76) famously—or perhaps infamously—attacked what they call the "adaptationist programme," the view that natural selection is "so powerful and the constraints upon it so few that direct production of adaptation through its operation becomes the primary cause of nearly all organic form, function and behavior" (see also Lewontin 1978, 1979). The adaptationist thought is that all traits are adaptations and that they are optimal, or at least nearly so (see Richardson 2003c). In place of a complacent commitment to adaptation as *the* explanation of organic form, Gould and Lewontin argue for what they describe as a more pluralistic approach to evolutionary biology, in which a wider array of alternatives to natural selection needs to be systematically considered. Showing that some trait is the product of natural selection, they contend, requires eliminating alternative explanations, or at least showing that they are less likely than adaptive explanations. The methodological complaint is that "adaptationists" do not seriously consider alternatives to adaptive explanations. So, they suggest, an adaptationist who offers one adaptive explanation for a trait will continue to assume the trait is an adaptation even though the original adaptive explanation might be disproven. Adaptation as a hypothesis wins by default rather than as an empirical result. It is important to understand that this is not a claim that adaptive explanations are "unfalsifiable," but that methodologically, the idea that a trait is an adaptation is often not tested, even though specific adaptive claims can be.

It is easy to be confused by the argument. On the one hand, it's tempting to infer that Gould and Lewontin are arguing that such speculations are "unscientific," in that adaptationism is unfalsifiable in principle (see, e.g., Buller 2005, 87). Sometimes they are taken to be objecting to an empirical claim concerning the ubiquity of adaptations rather than a methodological claim concerning the way we approach evolutionary explanations (see Sterelny and Griffiths 1999). That thought needs to confront the fact that Gould and Lewontin (1979, 589) explicitly allow that natural selection is the "most important of evolutionary mechanisms." It also needs to confront the fact that much of Lewontin's scientific work has been focused on the role of adaptation.

I think a consistent reading of the paper must treat their challenge as a challenge to adaptationist *methodology*. Others clearly defend adaptationism as a methodological view (e.g., Alcock 2001). John Maynard Smith (1978, 1987), for example, suggests that the goal of evolutionary analysis is not to test adaptation *per se*, but the factors governing and restricting adaptive evolution. The goal is to discover the constraints on adaptation, and not at all to test whether any particular trait is adaptive. (For more discussion, see Richardson 2003c.) The former, Popperian thought, that Gould and Lewontin claim that adaptive explanations are unfalsifiable, and so unscientific, is simply untenable. After all, Gould is a paleontologist, happy to infer past causes from current pattern. His work on the causes of coiling patterns in snails hardly displays an antagonism toward inferring history. Lewontin has done foundational work using empirical evidence to distinguish selection from drift as evolutionary causes. Certainly their point could not be that historical or evolutionary claims are unfalsifiable or unscientific. A more subtle treatment of the issues is necessary, although both advocates and detractors of evolutionary psychology resist the invitation (see, e.g., Buller 2005, chap. 3; Alcock 2001; Dennett 1995).

Such at least is the brief against "adaptationism." Gould and Lewontin go on to offer a variety of alternatives to natural selection, including genetic drift, genetic interaction (pleiotropy), and allometry, but the image of the spandrels of San Marco suggests strongly the role that developmental or structural constraints might play in evolutionary processes (see Gould 1977; Alberch 1982; Maynard Smith et al. 1985). An *evolutionary constraint* is a limitation on the course or outcome of evolution. An evolutionary constraint thus guides or restricts the range of evolutionary outcomes. A *developmental* constraint would limit the range or quality of outcomes based on developmental mechanisms. So if we look to the tetrapod limb, we find a characteristic structure and development. There is a common form, beginning with five digits and a "wrist." It can be warped and molded in various ways, and digits can be lost (as in *Archaeopteryx*), but there is a recognizable similarity of form. This structure is certainly "adaptive" in the sense that it can be modified to fit various forms of life. In one manifestation, it is a human hand; in another, it is a horse's hoof; in yet another, it becomes a bat's wing. And of course, it becomes *Archaeopteryx*'s foot, with a loss of digits and a reversed hallux. The focus on developmental constraints suggests that there are limitations on the form, across these various modifications, and that the form itself does not get an adaptive explanation. It is, instead, the *Bauplan* or blueprint that is modified as it shifts from humans, to horses, to bats, and *Archaeopteryx*. Thinking of these factors as *constraints* unfortunately underestimates their role in generating, rather than just limiting, novelty (see Grantham 2004). Developmental contraints are not merely limits on selection; they also serve to channel the direction of selection by channeling variation.

The architectural analogy should be a familiar one. Gould and Lewontin describe the central dome of St. Mark's Cathedral in Venice. The domed ceiling is mounted on arches, and where the dome meets the arches, there are "spandrels," triangular spaces formed at the intersection. Gould and Lewontin say that, in spite of the ornate use of the spandrels in religious symbolism, these spandrels are merely the consequences of mounting a dome on pillars. They say:

> The design is so elaborate, harmonious, and purposeful that we are tempted to view it as the starting point of any analysis, as the cause in some sense of the surrounding architecture. But this would invert the proper path of analysis. The system begins with an architectural constraint: the necessary four spandrels and their tapering triangular form. They provide a space in which the mosaicists worked; they set the quadripartite symmetry of the dome above. (Gould and Lewontin 1979, 582)

Spandrels, apparently adaptive structures that are in fact byproducts of other design choices, are direct challenges to Dawkins' gambit. The identification of spandrels as byproducts, the result of architectural constraints, is meant to be suggestive of the nineteenth-century idea that there are fundamental *Baupläne* (architectural building plans, or blueprints) characterizing animal form.[7] The fundamental and interesting thought is that just as spandrels are not design features but byproducts due to architectural constraints, so many apparently adaptive features in organisms may be not adaptations but byproducts due to developmental or phylogenetic constraints.

The attraction of the idea is not difficult to see. The construction of a phenotype from a genotype is a complex affair, and the idea that the phenotype could be indefinitely molded to meet the demands of the environment is certainly unrealistic. The case of the pentadactyl (five-digited) hand is a case in point, since the fundamental structure is certainly not an adaptation for grasping. The human hand is, in Darwin's word, a "contrivance." Even at the genetic level, pleiotropy and epistasis guarantee that there are nonlinear and nonadditive interactions among genes. That is, a change in one gene can, and often does, have dramatic and unexpected effects, and those effects can be amplified. The most dramatic and unfortunate examples of this are connected with deleterious genes, such as cystic fibrosis, a human mutation that affects a wide range of features. Moreover, there are dependencies in ontogeny that entail that characters cannot be readily decoupled (see, e.g., Schank and Wimsatt 1988; Wimsatt and Schank 1988; Wimsatt 2007): some structures cannot develop without precursors, and so even if those precursors have no later function, they can be retained later in development. These precursors are

Figure 2.5
The representation of spandrels in St. Mark's Cathedral. Just left of center, the spandrel rises in a triangular shape in the image. That is the surface roughly perpendicular to the line of sight. On either side of the spandrel are the arched tops to a square pillar. They lie at right angles to one another, along the midline of the "spandrel." From http://www.mrrena.com/images/fig1.jpg/.

developmentally required as a kind of substrate for other structures with adaptive effects. Moreover, suites of characters evolve in concert, in ways that are impossible to predict in the absence of specific developmental information. This virtually guarantees that there will be constraints that limit and perhaps guide evolutionary change. Still, it is one thing to notice that the idea is attractive; it is another to show that it is compelling. And it is still another to demonstrate its importance in a given case. Absent knowledge of *how* this happens, given that it does, or *why* it happens, the attraction is hardly compelling. The recent sophistication of work on development and evolution—"evo-devo" among its devotees—at least promises to give us some traction on the issues. There are, it turns out, specific genes that affect leaf and flower development in higher plants, and others that influence limb development. Unfortunately, following this out would take us very far afield (see Raff 1996 for a biological assessment and Burian 2004 for a philosophical peek at the issues).

The methodological indictment Gould and Lewontin offer of adaptationism is nonetheless not just that it ignores development. Instead, Gould and Lewontin charge that the "research programme" embodied in adaptationism fails to acknowledge that there are viable alternatives to adaptation in the evolutionary process. As a consequence, adaptationism per se stands as an untested claim. They complain that among evolutionists, adaptation is preserved at all costs. It might be tested in a particular case, but if one adaptive scenario is disconfirmed, it is simply replaced by another adaptive scenario. So although particular adaptive hypotheses may be evaluated, and even rejected, adaptationism is untested: adaptation is preserved even in the face of the disconfirmation of any number of adaptive hypotheses.[8]

In the early decades of the twentieth century, it was common to acknowledge alternatives to natural selection. These included some now discredited views, including the inheritance of acquired characteristics. (Often this is called "Lamarckism," despite the fact that Darwin was in this sense a Lamarckian and that it was not Lamarck's fundamental mechanism for evolution.) However, other nonadaptive factors were often acknowledged, including geographical variation and polymorphism within species; but toward the middle of the century, these factors tended to be dismissed. Gould describes this as the "hardening of the synthesis" (see Gould 1983). His idea, which historically is unimpeachable, is that, although the early decades of the twentieth century acknowledged many agents of evolution, by the middle of the century, the tendency was to acknowledge adaptation and natural selection as the prime movers of evolutionary change.

Gould is right about the history of the idea. It is, of course, a quite different question about what actually are the causes of evolutionary change. My own suspicion is that nonadaptive factors feature importantly. That is not a view lacking evidence. Others clearly think that *only* adaptation is an important factor. Helena Cronin (2005, 25) says, in a eulogy to G. C. Williams, "Only through adaptations were environments constructed, and only through understanding adaptations can we reconstruct them." Whichever view is right, Cronin's views could hardly capture the idea of reverse engineering more elegantly. These sorts of issues form the leitmotif of much of what follows. I am convinced, with Cronin, that adaptation is central to the history of life; but I doubt it is the whole story.

There are in fact many alternatives to adaptive explanations that are fully evolutionary. Chance plays a significant role in evolutionary processes. There are interactions among traits and genes that influence evolutionary change. There are effects due to the constraints of growth, what is called *allometry*. Every species is subject to a variety of developmental constraints. Thus it is unreasonable to insist that any trait *must* be an adaptation. This is a hypothesis to be tested. It is the tribunal of experience that must decide the matter.

Gould and Lewontin are often taken, not without cause, to be promoting a more ambitious agenda, but the reliance on optimality and reverse engineering in the analysis of adaptation is perhaps the clearest example of the sort of research they deplore. They claim that the "adaptationist programme" typically involves two steps. First, an organism is "atomized" into traits that are explained as independent structures optimally designed by natural selection. This amounts to assuming that genetic variation is both extensive and additive and that, as a consequence, phenotypes are almost indefinitely malleable. Without that assumption, we could not expect optimal design. If there is this broad adaptive potential and extensive variation, then organisms could be expected to be at least nearly optimal forms. Second, since assuming that traits are genuinely independent is generally unrealistic, and since that is widely known, interaction effects are incorporated as trade-offs. Competing demands on organismal design are acknowledged, but without compromising the commitment to optimal design. The focus on current function and the use of abstract design criteria, they contend, obscure the role of history.

Evolutionary psychologists are certainly aware of the criticism of adaptationism, even if they are often dismissive and occasionally abusive. Buss and his collaborators (1998) embrace the idea, at least in principle, that there can be many factors affecting evolution. Among the factors they acknowledge are developmental factors and environmental effects. There may be crucial keys that initiate developmental responses; to use their own examples, experience in committed relationships might cause jealous responses, and lack of invested parents may promote short-term mating strategies (Buss et al. 1998, 535–536). They say, very generally,

the evolutionary process produces three products: naturally selected features (adaptations), by-products of naturally selected features, and a residue of noise. In principle, the component parts of a species can be analyzed, and empirical studies can be conducted to determine which of these parts are adaptations, which are by-products, and which represent noise. Evolutionary scientists differ in their estimates of the relative sizes of these three categories of products. (Ibid., 537)

Buss and his coauthors recognize that optimality is not readily achieved: there may be time lags between environmental change and adaptive response;

historical effects can leave populations "trapped" on local optima; there may be a lack of the requisite variation; and often trade-offs limit what can be achieved. "Adaptations are not," they conclude, "optimally designed mechanisms" (ibid., 539).[9] In principle, no doubt, we can disentangle these various causes. In the end, they insist, the crucial questions are whether hypotheses are adequately formulated and consistent with the available empirical data. It is hard to imagine how anyone could dissent from this modest conclusion. The more interesting question is whether their methodology belies their modesty. It is one thing to formulate a hypothesis consistent with the available data. It is another to show that a hypothesis is one we should accept as true. In practice, the answers may be elusive.

Paul Griffiths usefully suggests that the issues surrounding adaptationism and its critics should be seen as concerned with hypothesis testing and evaluation. I agree. As Griffiths (1996, 524) explains, "The anti-adaptationist critique of 'reverse engineering' is not a rejection of adaptive explanation. It is a recognition that the adaptive processes and their results do not correspond one-to-one." So whereas Dennett, joined by Cosmides and Tooby, finds reverse engineering essential to good Darwinism, Gould and Lewontin see it as inimical to it. Griffiths sees largely an empirical problem, as do I. All parties should agree on the key question, though, and that is whether there is sufficient empirical warrant to support the evolutionary claims, or whether the methodology makes a serious empirical inquiry impossible. That is the question I will focus on in the remainder of this chapter, particularly with respect to reverse engineering insofar as it aims to provide evidence concerning historical function and biological adaptation.

4 The "Dangerous Passion"

It is useful at this point to illustrate, in a bit of detail, how reverse engineering actually functions in the literature within evolutionary psychology. Many examples are possible; I've already mentioned some. The key to reverse engineering is that we begin with an analysis of the behavior. This is, of course, the very business of psychologists, so this is their home ground. My primary question, though, is not how well these hypotheses fare as psychological claims, but how they fare as *evolutionary* claims: how the psychological data bear on evolutionary claims, and how evolutionary theory bears on the psychological claims. Our initial question is how behavioral or psychological data bear on evolutionary hypotheses and what substantive gain there is to psychology from evolutionary biology. That is, the ultimate question is how evolutionary psychology fares and functions *as* a piece of evolutionary theory.

David Buss's *The Dangerous Passion* (2000) can serve as a reasonable illustration of the approach. Buss is an important representative of evolutionary psychologists. His work is provocative, well regarded, and widely cited. *The Dangerous Passion* is in many ways an elaboration of Buss's *The Evolution of Desire* (1994), and related work (e.g., Buss, Larsen, and Westen 1996; Buss et al. 1992; Buss and Shackelford 1997; and Buss et al. 1999; cf. Daly, Wilson, and Weghorst 1982). This research is impressive in its scope and widely applauded among evolutionary psychologists. The concern in *The Dangerous Passion* is with jealousy—certainly a common and familiar enough human emotion—and its evolutionary significance. Buss (2000, 5) is clear that the project relies on the idea that jealousy is an adaptation:

Jealousy, according to this theory, is an adaptation. An adaptation, in the parlance of evolutionary psychology, is an evolved solution to a recurrent problem of survival or reproduction. Humans, for example, have evolved food preferences for sugar, fat, and protein that are adaptive solutions to the survival problem of food selection. We have evolved specialized fears of snakes, spiders and strangers that are adaptive solutions to ancestral problems inflicted by dangerous species, including ourselves. We have evolved specialized preferences for certain qualities in potential mates, which hope to solve the problems posed by reproduction.

A large part of Buss's work draws on differences between males and females concerning reproductive strategies. This is, of course, a prevalent theme among evolutionary psychologists.

The key evolutionary ideas have their roots in earlier research by Robert Trivers (1972) on parental investment. Broadly, females—including many mammals, but specifically human females—tend to be more heavily invested in offspring than are their male counterparts. Among mammals, this is expressed most obviously in internal gestation, parturition, and lactation. This initial asymmetry in investment is the key to Trivers's models. Both males and females generally need reproductive partners, but the problems this presents for males and females are very different. Since fertilization is internal, at least among mammals, the key problem for the male is uncertainty concerning paternity. For the male, sexual infidelity in the female would compromise his reproductive interests, making it more likely that he would be cuckolded and raise someone else's offspring as his own. Females have no such concern. There is, with mammals, including humans, no doubt about maternity. The female's problem is, instead, ensuring the commitment of her reproductive partner to her and her children. Lacking any doubt over maternity, the key concern is over emotional commitment. Jealousy, as a human emotion, is supposedly a reaction to these different evolutionary problems. Moreover, since the evolutionary "problems" faced by our male and female ancestors were different, the current profile of jealousy should be different between males and females too. Buss tells us that it is indeed different. To put it a bit too starkly, but only a bit too starkly, he contends that jealousy in men is invoked by sexual infidelity, whereas in women it is invoked by emotional infidelity. "Jealousy," Buss (2000, 5) tells us, "is not a sign of immaturity, but rather a supremely important passion that helped our ancestors, and most likely continues to help us today, to cope with a host of real reproductive threats."

Buss (ibid., 10) is likewise clear about what we should expect in light of the significance of the issues:

From an evolutionary perspective, no single decision is more important than the choice of a mate. That single fork in the road determines one's ultimate reproductive fate. More than any other domain, therefore, we expect evolution to produce supremely rational mechanisms of mate choice, rational in the sense that they lead to wise decisions rather than impetuous mistakes.

Love also provides an inroad to understanding our emotional structure. From an evolutionary perspective, love serves as a bond among reproductive partners. Again the key, for the evolutionary psychologist, is the difference in reproductive interests between males and females. What should we expect from these differences that would "produce supremely rational mechanisms of mate choice"? Since the "problem" for females is providing resources for their offspring, what is valued is whatever features reflect success and status: industriousness, intelligence, and creativity, for example. Males, on the other hand, emphasize the virtues connected with fertility since that is likely to improve their offspring: youth, health, and physical appearance loom large. Likewise, in *The Evolution of Desire*, Buss (1994, chaps. 2 and 3) contends that women want men with economic resources, social status, maturity, ambition, industriousness, dependability, stability, intelligence, compatibility, size, strength, health, and commitment. Men, in contrast, want youth, beauty, and good body shape. Again to put it a bit too starkly, but again only a bit too starkly, Buss concludes that women want security and men want beauty.

Buss's evidence for systematic sexual differences is occasionally anecdotal, but more often straightforwardly social-psychological in character.[10] So, for example, we are told that men are "more promiscuously inclined than women" (Buss 2000, 14); the evidence offered for this is a broad survey of men and women who differed in what they reported desiring. The evidence provided crosses many cultures and countries. Buss reports statistically significant differences in all but one case. Subjects were given questionnaires and asked whether they disagreed or agreed with a set of claims concerning, for example, how likely they would be to accept a sexual proposition from a stranger. Buss found differences depending on gender. So, for example, males typically said they desired having eight sexual partners over the next three years, whereas females typically reported wanting only one or two. Similarly, we are told that males and females differ in what they report would trigger a jealous response. This is the contribution Buss thinks especially significant to evolutionary psychology, apparently because it is inspired by Trivers's model for parental competition. Given the costs of failed paternity to males, in terms of wasted effort and opportunity, Buss (ibid., 52) says "we expect that evolution would have forged powerful defenses to prevent incurring them. Jealousy is the best candidate." Likewise, females risk losing commitment, and the diversion of her partner's resources to another: "The most reliable indicator that a man would divert his investment was not in having sex with another woman *per se*, but rather in becoming *emotionally involved* with another woman" (ibid., 53). Again using surveys, he determines that women more often say they would find sex without commitment "unsatisfying," and that they would find it "hard" to have sex with someone they did not love; men did not report those sentiments, saying they would be satisfied by sex without commitment, and that it would be easy to have sex with someone they did not love. Buss also found differences in what these subjects said would provoke jealousy. He provided various scenarios and asked them whether they would find it more difficult to forgive a partner who had had sexual intercourse with a former lover, or one who had rekindled the emotional connection with a former lover. He found that 67 percent of the men reported that they would find the sexual transgression more difficult, contrasted with 44 percent for the women in the study. More direct behavioral measures indicated that men would be more distressed by ideas of sexual infidelity, whereas women displayed more distress over emotional infidelity, though again their reactions were to imagined scenarios. Buss (ibid., 55) concludes:

Men and women, in short, differ in their attitudes about the role of emotional involvement in sex. Most women, even those who end up having a number of casual sexual encounters, want some kind of emotional involvement. Most men, in contrast, have less difficulty having sex without emotional involvement.

Buss cites evidence that these differences in attitude are reliable across cultures: men's jealousy is sensitive to sexual infidelity, whereas women's focuses more on emotional betrayal. He concludes that these differences are likely to be "universal sex differences" (ibid., 60).

I do not find the sort of evidence offered at all compelling, though, as I've said, my primary concern is not with the character of the psychological evidence. My concerns are not distant from those expressed by Kitcher and

Vickers (2002). They wonder whether Buss measures actual preferences, whether those preferences are stable, and crucially whether *any* of these preferences feature in actual choices of mates. They conclude, after negative assessments on each point, that Buss's view suffers from "psychological poverty" (Kitcher and Vickers 2002, 338ff.; cf. Buss 2000, chaps. 5 and 6). At the first level, in sexual surveys, there are pervasive problems in getting reliable reports. Men often tend to overreport whereas women to underreport sexual encounters. Buss (2000, 132ff.) even claims in *The Dangerous Passion* that men have more affairs than women; presumably, the data concern heterosexual encounters. In *The Evolution of Desire*, Buss (1994, 80) cites the Kinsey report to the effect that 50 percent of men, but only 26 percent of women, had extramarital affairs. Again, the point presumably concerns heterosexual encounters. This is, to say the least, unlikely. It doesn't matter whether what is reported is the net number of encounters or the average number of encounters. In the Kinsey report, it was used as evidence that the reports (of affairs) are unreliable, but it is not so used by Buss. He is curiously unconcerned over how it could be true. Indeed he draws the unlikely conclusion that "all studies show sex differences in the incidence and frequency of affairs, with more men having affairs more often and with more partners than women" (ibid., 81). At the second level, we humans are notoriously unreliable as predictors of our own behavior. What we would predict of ourselves, and our reactions, is often very different from what our reactions actually are. This is very clear in the literature from social psychology. Indeed, what are called the "intuitive" predictors of our own behavior, or that of others, have an unforgivably low predictive power concerning actual behavior (see Ross and Nisbett 1991). We tend, among other things, to underestimate the influence of the "situation" and overestimate the influence of "character," all the time with distressingly low predictive success. At either of these levels, Buss's work, as psychological work, is problematic. However, that is not my focus. My concern, as I've said, is not centrally with the psychological work per se. I want to focus on it as a piece of *evolutionary* psychology, and not as evolutionary *psychology*.

Aside from the problems with the data, we should ask: How does this help us understand our evolutionary heritage? Perhaps better, we should ask: How would this help us understand our evolutionary heritage, assuming the data were probative? Here is the way the evolutionary argument is evidently supposed to work. We begin by assessing differences between men and women in terms of their sexual attitudes. Let's suppose, contrary to fact, that we actually have reasonable gauges of these differences. Buss uses the evidence to conclude that there are sexual differences, notes the consistency of this with some evolutionary models, and infers that our ancestors not only actually behaved in a way that reflects the differences in our attitudes, but that there were selective pressures to so behave. These differences in attitude are supposed to reflect some deeper underlying differences in mating strategies. The mating strategies are taken in turn to reflect some fundamental biological imperative. This argument is put forth without citing evidence concerning, say, group structure, which would certainly be relevant. It is put forth without evidence concerning mating structures, which would certainly be relevant. It is put forth without so much as information concerning group size. It is put forth without information concerning similarities between ancestral and current group structures, mating structures, or group sizes. It is, of course, put forth without evidence concerning actual mating behavior, or the differences supposed to exist in ancestral groups. From all this, and less, we infer that there is an evolutionary and adaptive explanation of the differences between the sexes. That conclusion, in turn, allows Buss to interpret the evidence in a way that he thinks somehow makes evolutionary sense. If this is the way the argument works, then it is a case of reverse engineering. But it is not a very compelling case of reverse engineering.

Box 2.1
Parental investment

Parental investment is the contribution a parent makes to offspring that increases their offspring's chances of surviving and reproducing. Parental investment indirectly contributes to the fitness of the parent. This comes with a cost since any contributions to one offspring will reduce the contributions that might be made to another offspring, and may even reduce parental survival. We can think of the individual as having a limited set of resources to use in raising offspring. There can be asymmetries in the relative investments. In the typical case, the differences in reproductive metabolism suggest the female begins with more investment in an offspring than does her male partner. When there is internal gestation, the female has a commitment to the offspring of a sort the male does not. In Trivers's formulation, this makes the female the "limiting resource." This means that, in theory, an individual male can sire more children than can an individual female—though of course, that cannot translate into any average difference between the sexes. One option for the female is to require some additional male investment (e.g., in nest building) prior to sexual contact, or somehow to increase the commitment of the male. At that point, there will be more symmetry in "investment," which should lead to a more equal subsequent investment in the offspring. If the male is interested only in maximizing his number of offspring, he will have an interest in reducing his average contribution to offspring. The result is a kind of conflict of interests. The original work was done by Robert Trivers (1972) as an elaboration of Darwin's work on sexual selection. This has been modeled using game theory, which supports the idea that female strategies can increase the "fidelity" of males (cf. Dawkins 1976; Ridley 1993). Colorfully, males and females are thought of as engaged in a game in which the female seeks increased paternal investment and commitment while the male seeks decreased paternal investment and commitment.

5   Varieties of Design

I've observed that there are significant differences of opinion concerning how reasonable it is to use reverse engineering that are quite apart from any issues specifically concerning evolutionary psychology. Dennett is a fan of reverse engineering; Gould and Lewontin are not. If we see some systematic difference between men and women, in terms of what makes them jealous, or in terms of their tendencies to violence in response to infidelity, one key question we face is whether this supports an evolutionary, and adaptive, explanation. Buss's explanations are relatively specific, depending on differences between the sexes in terms of reproductive strategy. They are at least consistent with some evolutionary models, though mere consistency is hardly a difficult standard to meet. Just about anything is consistent with some evolutionary model or other. The working assumption is evidently that human social structures in our ancestral conditions were ones in which these differences in reproductive strategy would lead to the sorts of patterns he describes. This style of reasoning strategy is the one dismissed by Gould and Lewontin as untenable, a fabrication of "just so" stories without empirical support. We have seen that others, such as Dennett, think of this as the key to good Darwinian thinking. The responses to engineering design arguments have been nearly as numerous as the respondents. Here is what G. C. Williams (1992, 40) says, even in the context of emphasizing how difficult it is to show that a given trait is an adaptation:

Adaptation is demonstrated by observed conformity to *a priori* design specifications. This is the main method used by Galen and Paley . . . and recently advocated by Thornhill (1990). The hand is an adaptation for manipulation because it conforms in many ways to what an engineer would expect, *a priori*, of manipulative machinery; the eye is an optical instrument because it conforms to expectations for an optical instrument.

We find Steven Vogel (1998, 4) offering this somewhat more limited approval, though he is not focused directly on evolutionary questions as much as questions of current function:

much of the design of organisms reflects the inescapable properties of the physical world in which life has evolved, with consequences deriving from both constraints and opportunities.

These sentiments contrast sharply with the verdicts offered, for example, by Michael Ghiselin (1983, 363), who echoes Gould and Lewontin:

Panglossianism is bad because it asks the wrong question, namely, What is good? . . . The alternative is to reject such teleology altogether. Instead of asking, What is good?

Ask What has happened? The new question does everything we could expect the old one to do, and a lot more besides.

How one understands appeals to engineering design has distinctive consequences for the understanding of adaptation. If engineering design is a good method for uncovering adaptations, that seems to favor the position of Dennett, Cosmides and Tooby, and Buss. If engineering design is not, then that is support for Gould and Lewontin.

The divergence of opinion should give us some pause. How could there be such a wide divergence of opinion? Why would otherwise capable evolutionary theorists differ so widely over the method? I think it is useful to distinguish a number of applications of engineering design in the understanding of adaptation. I will offer a moderately simplified scheme, with some biological illustrations. These cases differ strikingly in how compelling they are, and that may in turn explain the divergence of opinion. The first thing to notice is that we can begin in different places. On the one hand, we can begin with an analysis of environmental structure, looking to the demands the environment places on the organism with the goal of inferring organismic function. This, as I've said, is often called "adaptive thinking." We look to the "problems" the environment poses and infer what the "solutions" should look like. Such adaptive thinking can and does involve significant appeal to design. On the other hand, we can begin with an analysis of organismic structure, looking to the structure of the organism to determine the organismic function, and from that infer the environmental demands. We look to "solutions" and infer what the problem must have been. This is Beatty's approach, with which we began the chapter. The behavior of an organism, no less than its structural features, is a reasonable target for either of these approaches. Indeed, Beatty begins with structure, but the issue is behavior. There is an equally fundamental difference in approach that is independent of this first distinction. Any design model will incorporate a variety of constraints. There are very different roles for the constraints that give substance to these models (see Amundson 1994, 1996; Amundson and Lauder 1994). The key issue is whether the constraints defining what counts as optimal form or behavior are *prior* to the assessment of fit between form and function or whether the constraints are inferred from, and *posterior* to, the assessment of fit between form and function. That is, we can begin with the constraints in place, or infer the character of the constraints from other factors. If we begin with the constraints, information concerning those constraints can be incorporated into evolutionary models as prior constraints, independently established, on the available range of biological form. It is useful to think of this as an *a priori* structuring of the "problem," to use Williams's language. Environmental information then gives a structure to the "problem," constraining the alternatives systematically without assuming that there is a fit between form and function.[11] On the other hand, we can develop design constraints *a posteriori*, after the fact, using the performance of the phenotype to determine the environmental factors that shaped that design. This is exemplified in the case I've already described above from John Beatty (1980). This type of approach is also the most ambitious case of "reverse engineering" and is Dennett's favored case.

This results in a four-way categorization, illustrated in figure 2.6. On each dimension, the differences are probably better thought of as matters of degree rather than kind. As a useful approximation, I'll treat them as if they are qualitative differences.

The issues are actually considerably more complex than this simple taxonomy allows. There is another dimension to the problems that ideally would be recognized. It would make our discussion much more complicated, and I'm not sure what the implications would be, although I don't think they would dramatically affect our conclusions. One dimension introduces depth to the understanding of organismic analysis; the other brings depth to the environment. George Lauder (1996) emphasizes that there is often no ready connection between an analysis of structure, behavior, and (physiological) function. He considers four taxa among salamanders: *Abbystoma*, *Cryptobranchus*, *Necturus*, and *Siren*. There are significant differences among the four groups. Considered from the point of view of structure, *Siren* and *Abbystoma* are indistinguishable, whereas *Necturus* and *Cryptobranchus* are differentiated. Behaviorally, *Abbystoma* and *Necturus* are indistinguishable, whereas *Siren*

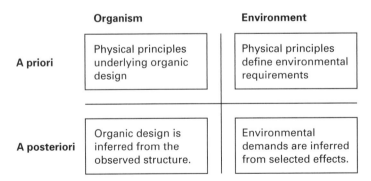

Figure 2.6
Four forms of engineering design. A four-way categorization of arguments from engineering design, depending on what considerations are used to generate conclusions concerning design. Constraints may be known independently, and hence count as *a priori*, or may be inferred, and hence count as *a posteriori*. The constraints, in turn, may depend on the structure of the organism, or the nature of the environment. Each of the four categories is illustrated in this chapter.

and *Cryptobranchus* are differentiated. And in terms of function, *Necturus* and *Cryptobranchus* are indistinguishable, whereas *Abbystoma* and *Siren* are differentiated. So even though *Siren* and *Abbystoma* are similar in morphology, they differ with respect to both behavior in feeding and physiological function. And even though *Abbystoma* and *Necturus* are morophologically similar, they are functionally distinct. Lauder concludes from this that "It is clear that an analysis of any one level alone is an insufficient description of the design of the feeding system in salamanders, and that prediction of behavior or physiological function from structure alone in this case study is effectively impossible" (Lauder 1996, 70). Structure, behavior, or (physiological) function each could be a target for evolutionary explanation.

Similarly, Robert Brandon provides a much more complicated proposal for understanding the environment. Brandon's entry point is the recognition that natural selection and adaptedness are relative to a *common environment*. Brandon offers a simple argument for this conclusion. Given two different plants, with dispersal of seeds by wind, a disproportionate number of one type may by chance land on fertile ground. This will result in a differential increase of that plant. This, he says, is *not* natural selection. The differences in realized fitness are not due to differences in adaptedness. Brandon (1990, 46–47) concludes that "in order to explain differences in realized fitness in terms of differences in adaptedness one must compare organisms in common environments." Brandon's actual example is simply of a chance effect in a heterogeneous environment, and that, as he correctly says, is not an instance of natural selection. It is not obvious to me that the example Brandon offers is one in which the organisms lack a common environment. Brandon is nonetheless correct that natural selection is not at work here. A simpler, but analogous, argument yields the same conclusion and supports the requirement for a

Table 2.1
Similarities among groups of salamanders

	Taxonomic Groups			
	Abbystoma	*Cryptobranchus*	*Necturus*	*Siren*
Structural	X	Y	Y	X
Behavioral	X	Y	X	Y
Functional	Y	X	X	Y

We begin with four groups of salamanders, and rate their relative similarity on structural, behavioral, and functional criteria. The several criteria group the taxa differently. Each of the four groups is indicated by similarity with an allied form. The four differ in how "similar" they count depending on the criteria that are deployed. For details see Lauder 1996.

common environment. A plant in Borneo might differ in any number of ways from another in Haiti, or they might be (genetically identical) clones; however, lacking a common environment, they simply cannot be compared with respect to fitness. Even though one may proliferate, and the other might decline in numbers and finally go extinct, this is not a difference in their relative adaptedness. Natural selection may be in operation, but there is no selection between these two forms. Population pressures may differ. A drought may cause a population crash on one island, a typhoon may sweep through another. There may be different competitors. There are doubtless many differences, but this is not natural selection. Any comparison of the *relative* adaptedness of these two plants is empty. Fitness is always a relative measure, after all. As Brandon says, though two plants in different environments may differ in reproductive success, the two seed types might not differ at all, and without a common environment, whatever differences there are between them should not be seen as due to natural selection promoting one over the other.

Natural selection is thus relative to a common environment, however exactly we should understand that. Brandon goes on to distinguish three concepts of the environment. The *external environment* is partitioned on the basis of independent physical or biotic factors, such as rainfall, temperature, food sources, or predators; any differences can be used to discriminate parts of the external environment. The *ecological environment* is partitioned on the basis of differences in the performance of organisms, and thus is dependent upon only those factors that affect fitness; the ideal measure of differences in the ecological environment is the performance of the same type across space or time. Finally, the *selective environment* is partitioned on the basis of relative differences in the performance of organisms, and thus is dependent upon only those factors that *differentially* affect fitness; the proper measure is the relative performance of different organisms within the same ecological environment.

These three ways of partitioning the environment do not always partition it in the same way, any more than the ways of understanding organisms partition them in the same way. Thus, if local variations in some bacterial concentrations in the soil do not affect adaptedness within a population of plants, then although they are part of the external environment, they are not part of the ecological environment for those plants. The bacterial levels are invisible to the plants. Furthermore, even if these bacterial levels affect fertility, they may have the same relative effect. This could be observed by comparing the reproductive output of genetically identical plants in different ecological environments. If these variations do affect the fertility of different plants, but do not change their relative fitness, then they are part of the ecological environment but not of the selective environment. This could be seen by comparing strains across environmental variation. If these variations affect the relative fitness of the plants, then these are differences in the selective environment as well. Brandon (1990, 66) says that from "the point of view of the theory of natural selection, the relevant environment is the selective environment." If we are to infer from environmental structure to organismal function, or vice versa, it is important to distinguish which sense of "environment" is being used.

The differences observed by Lauder and Brandon would give us a yet more complicated picture of the forms of engineering design. Instead of four possibilities, we actually have at least twelve, and probably a continuum of cases in three dimensions. It may be that acknowledging these differences would affect our verdicts on the quality of the inference, though I don't think this is so. They certainly make the inferences more complicated. I will ignore these additional dimensions and focus on the simpler four-way partition.

So let's look at some examples from each of the four quadrants in the more simplified four-way scheme. In the first quadrant, at the upper left, we begin with models that incorporate independent *a priori* constraints on organismic design. Vogel recognizes that the designs of nature are imperfect and incomplete. He nonetheless finds the assumption that there is a "decent fit between organism and habitat a useful working hypothesis" (Vogel 1988, 10). He presents a number of fascinating cases. One vivid illustration of physics at work in the biological world derives from Bernoulli's principle (after Daniel Bernoulli, 1700–1782), which explains why airplane wings give lift. The principle essentially says that the pressure from a fluid decreases as the rate of flow of the fluid increases. So consider a simple plane surface with a hole and fluid flowing horizontally (it could be liquid or gas, but it must be a fluid with relatively low viscosity). As the fluid moves more rapidly across the opening, the pressure from the fluid will decrease. If we start with the pressure at equilibrium above and below the surface, the result would be to draw the fluid from below to above at a rate proportional to the difference in rate of motion. As it turns out, we find the principle at work in prairie dog burrows. Burrows have more than one opening. Changing the shape and height of the openings brings Bernoulli's principle into play. Air flow will be more rapid for raised openings, and that will create a pressure differential, resulting in air motion through the burrows. A slight breeze will then draw air through the burrow. Showing exactly how this works depends on burrow lengths, depths, and the distance between openings. Prairie dogs work adeptly to maintain a difference between the heights of openings in such a way as to maintain circulation. As a result, air within the burrow is replenished, and the animals are not asphyxiated in the bottom of the burrows. Of course, prairie dogs did not read

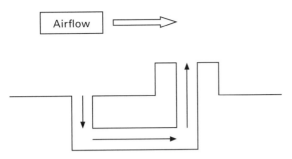

Figure 2.7
Bernoulli's principle. Given some airflow across a raised opening, with a slower flow across a connected lower opening, Bernoulli's principle tells us that air will be drawn through the tunnel since the airflow will cause a pressure gradient (as indicated by the arrows). Prairie dogs maintain elevated openings to promote such airflow through their tunnels. Sponges likewise use ambient flow to induce water through their central chamber, along with food.

Bernoulli, but in a sense they understand how the principle works. It does not take much brain power to use the principle, even if it takes a genius to articulate it. Sponges—not one of the more intelligent organisms—similarly use pressure differentials to move water through for filtration.

Here is another simple example, again from Vogel. Leaves are flat, and that results in a problem with bending. A planar structure is generally not rigid; at least, not without being very bulky, which imposes a number of collateral limitations. One solution is to impose "beams." The veins on leaves provide simple support structures that make them relatively rigid, as does introducing curvature along the midrib. The veins increase the effective thickness of leaves, with the result that tension can resist bending but without unnecessarily increasing mass. Curvature along the midrib results in a mildly concave curvature that also resists bending. The flexibility (which would be lost by merely increasing thickness) in turn allows some leaves to fold when the wind increases, reducing drag and damage. In cases such as these, the constraints on design derive from the physical demands, and we know not only in broad strokes but in detail what those physical demands are.

In the second quadrant, we have models in which there are *a priori* constraints imposed by environmental structure. These are especially elegant cases of engineering design arguments. I think these in fact are the most compelling cases for adaptation. In this use of engineering design, information concerning constraints is incorporated into models as constraints on the available range of biological form, specified prior to the assessment of adaptation. Thus, among the ubiquitous marine water striders, *Halobates* can move across the surface of seawater with amazing quickness and can also jump and land on

Figure 2.8
Halobates. It is sometimes known as the "Jesus bug" because it can walk on water. The total contact line of its feet is carefully matched to the surface tension of the medium. From http://www.zmuc.dk/entoweb/Halobates/images/HALOBA01.JPG/.

the surface of the water. There are two physical constraints that make this remarkable trick possible. First, like many insects, the water strider has a coating that makes it resistant to surface wetting. Second, there is a match between the total perimeter of its feet (which determines the contact line for the striders) and surface tension (which varies with salinity). Physics suffices to determine the surface tension, once salinity levels are known. With a mass of roughly 10 milligrams, the contact line would need to be roughly 1.3 millimeters total to move about and roughly 13 millimeters to sustain a jump (cf. Vogel 1988). *Halobates* meets the expectation of the physical model. It has specialized hairs (analogous to snow shoes) to increase its total contact, thus making both movement and jumping possible. In cases such as this, the constraints are easily defined beforehand, since they depend on physical parameters (body mass and surface tension). This *a priori* use of constraints is, in itself, not objectionable. It amounts to using environmental, developmental, or physical information to structure the "problem," constraining the alternatives systematically. The result is that we have principled grounds for evaluating the adequacy of the "design."

   It is, unfortunately, difficult to use these kinds of approach generally, even if it is illuminating when we can. George Lauder (1996, 71) observes:

The claim that relevant design criteria can be specified *a priori* to allow the analysis of biological design amounts to a claim that we can specify in advance the problem or problems that the design is supposed to solve. Although it is almost always possible to specify *some* design criterion, the more complex the design, the less likely it is that we will be able to determine what the relevant performance and mechanical functions are that any given structure needs to solve. And furthermore the less likely it is that we will be able to meaningfully weigh alternative performance goals.

The difficulty is simply that *a priori* design constraints are often not specified and, in many cases, cannot be specified before the fact. It is, in fact, in the less complex cases that it is easiest to specify the design criteria, as in the case of *Halobates* or leaf structure; as the structures under consideration get more complex, the chances of knowing the relevant design parameters are reduced. As we saw in the case of *Archaeopteryx*, the relevant design criteria are many in assessing the capacity for flight; showing that *Archaeopteryx* was actually capable of flight would require evaluating the entire package. As we saw, the most telling features are negative, suggesting *Archaeopteryx* did not fly. Given, for example, its lack of a sternum, poor wing shape, and foot design, it seems less than ideally designed for flight.

In the quadrant on the lower right, we argue *a posteriori* for constraints on environmental structure. The point of MacArthur's (1957, 1960) classic work on the abundance of species in the "broken stick" model was that the existing data on relative abundance of species fit better with a model based on nonoverlapping niches. These data, MacArthur concluded, fit better with an analysis that assumes competitive exclusion than with an analysis that assumes species distribution patterns are caused by, say, abiotic factors. This is evidence for the evolutionary explanation of nonoverlapping niches. The analysis of pattern is, for MacArthur, a vehicle for understanding evolutionary history. MacArthur found a rough qualitative fit between the observed distribution of species and a competitive model. The fit is by no means perfect, though. (Of course, qualitative fits make less than compelling cases.) Common species are more abundant than the model predicted, and rare species are less abundant than predicted. MacArthur pointed out that if the environment is heterogeneous, then it is possible to improve the fit between data and model. He goes on, in a remarkable piece of reasoning, to reverse the dependence: "The divergence from the ideal curve may, in fact, be regarded on this hypothesis as a measure of environmental heterogeneity. Experimentally, for bird communities, this appears to explain most of the 'steep' curves" (MacArthur 1957, 293).[12] This is a standard strategy in handling a mismatch between model and prediction, in order to explain apparent lack of optimal design (see Kingsland 1985): a deviation from initial predictions is explained by superimposing a second application of the same optimization model, taking up the slack in the fit by *assuming* optimal design. In such *a posteriori* applications, the degree of mismatch with the predicted optimum is used as a measure of the significance of the constraint. There is little that is obviously objectionable in this sort of reasoning, so long as we understand what is being evaluated. In MacArthur's case, what is empirically derived is a parameter indicating environmental heterogeneity. We no longer test for adaptation or attempt to support empirically even a specific adaptive hypothesis, and we certainly do not demonstrate optimality of design; rather, we assume it explicitly at the outset.[13]

Finally, in the quadrant on the lower left, we find *a posteriori* constraints on physiological design. Williams, for example, claims the human hand is a consequence of selection, recognizing that this depends on the sort of variation that was historically present and recognizing that there are alternative "designs" that might work equally well for grasping. This is a recent version of the problem of the vertebrate limb. Selection might explain, for example, the presence of five digits (even if six or four might result in a better-engineered tool for grasping) as a result of frequency-dependent selection. It might also explain quantitative characters, such as digit length. It can even explain why, when the functional demands change, as they did with the horse, or the bird, there was a loss of digits. Williams does not claim the pentadactyl structure actually was selected *for*, as far as I can see, and may intend the case primarily as illustrative. Lauder takes the example to task, as one for which there is too much freedom in defining the constraints on design. The hand is certainly a complex device, with twenty-six separate bones, including five metacarpals, fourteen phalangeal elements which constitute the fingers, and seven carpal bones. Additionally, there is a variety of associated nerves, blood vessels, and tendons. Structurally very complex, the hand is evidently designed for grasping. But which of these structures are adaptations for grasping? Are we supposed to think that the pentadactyl structure is an adaptation for grasping? Why, for that matter, are four fingers and a thumb better for grasping than three or five fingers and a thumb? And of course, there are many creatures with five digits that do not have our manipulative abilities. It is at least clear that much of the structure reflects ancient features common to vertebrate forelimbs. These are surely independent of the need for manipulation, from an evolutionary point of view.

I take it that we have no empirically grounded answers for such questions. As far as I can see, the pentadactyl hand, as such, is not an adaptation, and certainly not an adaptation for grasping. (Of course, there are modifications of the pentadactyl form that are adaptations for grasping, even if the pentadactyl hand is not.) As Lauder (1996, 75) says,

it is clear that the possession of independently mobile jointed elements ("fingers") is not a design component that could be linked to any specific function that is unique to the human hand: fingers are an ancient design feature of the vertebrate forelimb . . . and occur in many animals that do not have the manipulative abilities of the human hand.

Of course, the hand *is* used for manipulation. The ability to grasp and manipulate tools doubtless was *influenced* by natural selection; but saying how this is so in any informative way is just not that simple. The foot of *Archaeopteryx* was doubtless used for perching, even though it is not an adaptation *for* perching. The five-fingered structure of our hand supports human dexterity. It also allows for the ability to type. Beginning here leaves us woefully short of anything resembling a *principled* analysis of design.

6   Epicycles and Explanations in Evolutionary Psychology

Timothy Ketelaar and Bruce Ellis (2000) are anxious to defend evolutionary psychology from the charge that it is incapable of being falsified. Evidently, they think this is Gould's complaint about adaptationism. They also assume a commitment to reverse engineering. Here is what they say concerning the worries over falsification:

A major concern is the apparent willingness of evolutionary psychologists to generate a plethora of seemingly untestable and post hoc explanations for any given psychological phenomenon. (Ketelaar and Ellis 2000, 1)

Their counterpoint, in large part, is directed against the work of Gould and Lewontin, which they evidently take to imply that the "basic assumptions of modern evolutionary theory" are untestable and that the "specific evolutionary models and hypotheses that are drawn from these basic assumptions" are untestable (ibid., 2). As I've suggested, although there are passages in the "Spandrels" paper that raise the specter of falsifiability, the concerns of Gould and Lewontin are not actually Popperian. They are, at root, concerns about whether the methodology insulates the hypothesis of adaptation from empirical test, and not whether specific adaptationist claims are subject to test. Popperianism, with its emphasis on demarcating science and nonscience, could not be further from the issues at hand.

In the first half of the twentieth century, logical positivists were very interested in what they called a "demarcation criterion," capable of marking off scientific claims from merely "metaphysical" ones. Were there such a criterion, there would be some critical difference between scientific and unscientific claims. The thought many logical positivists had was that some claims in one way or another confront empirical issues. These are "cognitively meaningful" issues, the substance of science. Anything else had no cognitive significance. Karl Popper (1959) made a great deal of the (alleged) fact that positive claims cannot be conclusively verified, although they can be conclusively falsified. He supposed, as did many of his contemporaries, that scientific laws had the form of universal generalizations. Since these are unbounded, no number of positive instances could show that a scientific law is true. However, even one counterexample shows it is false. The corresponding suggestion, which Popper exploited throughout his career, is that scientific claims are distinguished from others by being "falsifiable." That does not mean they can be readily falsified, and of course it does not mean that they actually have falsifying instances; they need merely be falsifiable "in principle." He thus used *falsifiability* as a principle demarcating science from nonscience. He also thought that falsification offered a solution to the problem of induction—that the method of conjectures and refutations would lead to scientific progress where an emphasis on confirmation would not. Ketelaar and Ellis thus evidently take the issue to be whether evolutionary psychology is science or pseudo-science.

My own view is that this particular issue is not a central one and does not come close to the interesting questions. What is at stake is a much more interesting issue, and it is one on which I would certainly not expect agreement. Kitcher (1985, 61) tried to draw out the interesting issues this way in discussing whether evolutionary theory is defensible:

we need to do two things. First, we must show how a particular sequence of observational findings would have made it rational to abandon the theory. Second, we must examine the ways in which the actual evidence has confirmed evolutionary theory and attempt to understand which parts of the theory receive the strongest support.

Of course, Kitcher recognizes that evolutionary claims are often abandoned in the face of evidence. Suppose we are offered a specific "Darwinian history" (see Kitcher 2003)—that is, a specific explanation for some trait, an explanation consistent with Darwinism. As Kitcher suggests, there are at least two sources of methodological reticence concerning such histories: first, a favored Darwinian history could be defended no matter what empirical evidence is forthcoming; and second, even if a specific Darwinian history came to be abandoned in the face of empirical evidence, there would always be another Darwinian history to replace it. The first concerns whether a specific explanation can be disconfirmed, or could be retained in the face of prima facie contrary evidence. The second concerns whether Darwinism, broadly conceived, could be disconfirmed.

Kitcher observes that specific Darwinian histories are very much on a par with other empirical claims. Kitcher's own example is drawn from biogeography. Here's the case more or less as he offers it. There is a group of insectivorous mammals, tenrecs, which live on the island of Madagascar. There are roughly thirty species, and they range in size about as much as mice and rats.

They differ quite a lot among themselves behaviorally. Some burrow; some forage above ground; some climb; others are spectacular leapers; some are even aquatic, eating frogs and crustaceans rather than insects. They are often described as "primitive," although their evolutionary history is precisely as long as our own. Whatever "primitive" might mean, tenrecs have a number of features that would make them unlikely candidates in the struggle for survival. Their eyesight is poor; though they are mammals, they do not maintain a constant body temperature well; the young are immature; their reproductive and digestive tracts do not have distinct openings as is common among mammals. The question is how they came to be in Madagascar, and only there. The prevailing answer has much to recommend it. The ancestors of tenrecs arrived in Madagascar some sixty or seventy million years ago, about the end of the Cretaceous. Madagascar at that point was separating from Africa, with the expansion of the Mozambique Channel. Tenrecs were there. So were a number of other mammals, including some rodents, lemurs, and other even more exotic mammals. Many other mammalian forms are not found, including cats, canines, elephants, and others. Tenrecs, like Darwin's finches, lived long and prospered.

This is a sketch of a Darwinian history, though the full history is much more interesting and detailed. Tenrecs are an isolated group derived from an ancestral group in Africa. Could this Darwinian history have been falsified? Well, we might not find fossils where we expected they should be. Finding fossilized tenrecs in Tierra del Fuego would be a problem. We might find the only allied forms are quite distant from Madagascar. If the nearest relatives were native to Iceland, we would need a very different explanation for the presence of tenrecs in Madagascar. We might have found, based on geological evidence, that Madagascar did not split from Africa, but is moving toward the continent on a collision course. It turns out that this is not so. Had it turned out to be so, it would have told against the Darwinian history we now accept. As it turns out, the nearest relatives are mammals known as otter shrews, living in western and central Africa, and nothing close is present in Tierra del Fuego or Iceland. And though Madagascar was once part of Gondwanaland—the supercontinent that included Africa, India, South America, and Australia before they drifted apart—it seems it was at least nearer Africa than South America and much more distant from Iceland. It is a point of contention whether Madagascar split reasonably late from Africa, or whether it separated about the same time as others. In any case, Madagascar turns out to be very old, which is conducive to isolation and divergence. At some points, it was surely connected to Africa when the sea floor had not subsided much and the water levels declined. The reverse side of Kitcher's story should also be clear. There is an abundance of evidence supporting the dominant Darwinian history of tenrecs. There is not only the support of the proximity related species; there is also information from deep sea drilling and paleomagnetic data that is relevant to the claim that Madagascar split from Africa. There are other species, such as lemurs, present in Madagascar. Patterns of glaciation also help the story. And despite uncertainties associated with continental drift, even this fits the pattern we would expect.

Thus, tenrecs fit the Darwinian history well, and little counts against it. Kitcher (1985, 63) concludes from his discussion:

Darwinian histories are rationally abandoned when puzzles multiply at a faster rate than explanatory successes. . . . The sensible core of the idea that the individual accounts put forward by evolutionary theorists are falsifiable is the thesis that certain combinations of observational findings could generate a sufficient number of puzzles to make it unreasonable to persist in defending the accounts in question.

The modest moral Kitcher draws is just right. Specific Darwinian histories are subject to a variety of empirical tests and to broader empirical constraints. The history proposed for the tenrecs can be evaluated by looking for similarities to species across the Mozambique Channel and by turning to evidence for continental drift. The question remains whether anything could undercut the broader commitments. The same points recur: there *are* specific empirical constraints to be met if *any* Darwinian history is right. As Darwin himself saw, common descent requires some broad similarities among species. It is important that there are embryological similarities among living things. Among mammals, there are similarities in development. Among animals, there are similarities in specific genes, such as the hox gene. Animals even share a common genetic code with plants. These are explained by common descent. That is, they require a common evolutionary history. This doesn't tell us much about specific Darwinian histories, but it does tell us a lot about our world.

The same pair of questions applies to adaptive explanations as to "Darwinian histories." We can ask under what conditions specific adaptive hypotheses are abandoned or embraced in the face of empirical evidence. We can also ask under what conditions adaptation might be abandoned or embraced in the face of empirical evidence.

There is no serious question whether specific evolutionary explanations, and adaptive explanations, are testable. They are certainly falsifiable and empirical. They have in many cases been amply confirmed, and in some cases they have been falsified. Illustrations of adaptive explanations that have been confirmed, by a variety of evidence, are easy to come by (see, e.g., Endler 1986). To use a textbook example, the sickle cell gene ranges widely through native groups in central Africa, India, and Central America. In heterozygous form, it provides some immunity to malaria and was sustained by balancing selection, though our efforts at removing vectors (mosquitoes) have disrupted that balance. Similarly, the discovery in 1928 that penicillin is an antibiotic and its subsequent use in treating infections imposed simple, extreme, directional selection on bacterial pathogens. The result is, unfortunately, antibiotic-resistant strains of bacteria. *Cepaea nemoralis* is a snail common in Europe. Thrushes are fond of the snails, and the broken shells left behind offer a record of selection that varies with habitat and season. The result in this case is a shifting pattern of selection that results in a polymorphic population. All these are adaptations, shaped and maintained by natural selection. These explanations are relatively uncontroversial. They are all evolutionary; they all invoke natural selection; and they all enjoy considerable empirical support.

This hardly gets to the core issue, which is not whether specific adaptive explanations are sometimes abandoned or embraced in light of empirical results. The harder issue is the extent to which the hypothesis that some trait is an adaptation to something or other is subject to empirical test, and whether methodologically this general assumption should be insulated from test. This is the way Gould and Lewontin (1979, 587–588) make the point:

We would not object so strenuously to the adaptationist programme if its invocation, in any particular case, could lead in principle to its rejection for want of evidence. . . . Unfortunately, a common procedure among evolutionists does not allow such definable rejection for two reasons. First, the rejection of one adaptive story usually leads to its replacement by another, rather than to a suspicion that a different kind of explanation might be required. Since the range of adaptive stories is as wide as our minds are fertile, new stories can always be postulated . . . Secondly, the criteria for acceptance of a story are so loose that they may pass without proper confirmation. Often, evolutionists use *consistency* with natural selection as the sole criterion and consider their work done when they concoct a plausible story.

This certainly *can* be read as a complaint that "adaptive stories" are not falsifiable. That is not the most fruitful interpretation. We can profit from a non-Popperian rendering. There is at least some ambiguity whether the complaint is that some adaptive explanations are protected from falsification, or whether adaptation, per se, is thus protected. I do not believe for a moment that the argument from the "Spandrels" paper supports the conclusion that either of these claims is unfalsifiable in a Popperian sense. I also don't believe they were intended to enforce that conclusion, though I admit the rhetoric sometimes overreaches. The key point is that a methodology that insulates an explanatory scheme from confronting the data is a methodology that also makes it nearly impossible to know whether the explanation is the right one. Kitcher (1985, 232) puts the conclusion this way:

the correct position is that the successful pursuit of adaptationist hypotheses about traits of organisms already presupposes just that attention to rival possibilities that Gould and Lewontin urge upon their colleagues. Significant confirmation of adaptationist hypotheses is possible—but only if biologists are prepared to take seriously all the forms of evolutionary scenarios that they admit as possible and are prepared to undertake the investigations necessary for articulating claims about allometry, pleiotropy and so forth.

The positive moral is that we should proceed comparatively, asking whether adaptation is needed, or whether there are other equally plausible explanations. Some of the alternate explanations that Gould and Lewontin offer include the effects of pleiotropy, linkage, allometry, and developmental constraints. We could readily include drift, and more.

Gould and Lewontin simply do *not* claim (as Popper unfortunately once did) that evolutionary explanations are *unfalsifiable*. This would be nothing short of insane for an evolutionary biologist engaged primarily in the study of the importance of adaptation and drift in natural populations (Lewontin) or a paleontologist engaged in the history of life (Gould). They do complain that an *adaptationist* methodology can isolate adaptation from critical tests and that an *adaptationist* methodology can make nonadaptive causes invisible. Their issue is not with metaphysical claims—those beyond empirical test and refutation, which was Popper's preoccupation—but with methodological practices and how they affect scientific progress. It is not over what is untestable or unfalsifiable so much as what is untested and perhaps false. The key issue they raise, and which I echo in this volume, is not the falsifiability or unfalsifiability of evolutionary claims, or even of adaptive claims, but rather what is methodologically fruitful and what is methodologically barren. This is in fact not far from many of the key ideas Imre Lakatos pressed against Popper. In any case, Popperian methodology is not a relevant factor in Gould and Lewontin's critiques of adaptationism. It is also no part of mine.

In defending the status of evolutionary psychology, Ketelaar and Ellis appeal to a variety of cases, including that of sexual jealousy, discussed above. Their work is, as I've acknowledged, in many ways a broader philosophical project. They focus, particularly, on the use of competing hypotheses and the way hypotheses are modified within broader theoretical contexts. They observe that science typically proceeds under the umbrella of a set of assumptions that are not challenged, but assumed. These assumptions form a kind of framework within which research is conducted. Ketelaar and Ellis call this the "metatheory." It is, give or take a bit, what Thomas Kuhn called a "paradigm." Ketelaar and Ellis think of this as an improvement over Popperian thinking, and I agree. Naive falsificationism does not mark off science from nonscience and is not even an especially useful method for scientific discovery (see Bechtel and Richardson 1993, a book inspired in many ways by Lakatos). Ketelaar and Ellis favor an approach closer to that of Lakatos, one of Popper's students. Lakatos, like Kuhn, recognized the importance of theory and the role of theory in structuring experiment and observation. Ketelaar and Ellis (2000, 13) explain what they think is the crucial shift away from falsification this way:

> According to Lakatosian philosophy of science, then, the key scientific criteria for evaluating evolutionary psychology's guiding metatheory is not whether its core assumptions are false or not yet falsified, but rather how well the metatheory accommodates anomalies and whether it leads to fruitful new discoveries, explanations and avenues of research.

Lakatos distinguished the "hard core" of a theory from its midlevel applications, which he thought of as a "protective belt" that could be modified without affecting the hard core. His own exemplar was the development of Newtonian theory. The "hard core" of evolutionary theory from a Lakatosian perspective would be, at least, the results of the synthesis in the 1930s that wedded the theory of natural selection to Mendelian inheritance. The protective belt would be the applications of the theory, such as Trivers's work on parental investment, work on reciprocal altruism, and perhaps Hamilton's seminal work on kin selection. Lakatos thought theories were insulated by their "protective belt," involving various auxilliary hypotheses that could be modified without affecting the hard core. When these auxiliary assumptions are used to extend the core theory, Lakatos described the theory as progressive; when they are used primarily to adjust and modify the theory without extending it, the theory is degenerative. A good theory is progressive.

What Ketelaar and Ellis think of as the Lakatosian "hard core" of evolutionary theory—its metatheory—is thus secure, insofar as it amounts to a commitment to natural and sexual selection as central to evolution. In fact, they assume that the hard core metatheory is very limited, committed to a gene-based adaptationist theory (see Ketelaar and Ellis 2000, 4). If this is so, then the quarrel should be with what constitutes the relevant background, or metatheory. Gould and Lewontin are both evolutionary biologists, and the key issue they raise can be understood as one concerning what Gould called the "hardening of the synthesis," the exclusion of nonadaptive factors from the "hard core." At most, I think we should conclude that the core is unsettled. Alternatively, we might take the "hard core" to be something like what we see in contemporary works in evolutionary biology and population genetics. The problems Gould and Lewontin raise, then, are no longer preempted by the "hard core," but are issues over the range of application of evolutionary models. If we follow this line, as I do, then the problem is not over the metatheory but its application.

It should be clear that my own issues with evolutionary psychology are neither revolutionary nor reactionary. I am comfortable with evolution. I assume it. I am comfortable with natural selection. I assume it.[14] This by no means implies that all evolutionary explanations are testable, or that it is always practically feasible to evaluate the claim that some particular trait is the product of natural selection or that it is optimal. It certainly does not place the Popperian question of falsifiability at the center of the discussion.

A stringent adaptationist perspective assumes that complex traits are adaptations and that the postulation of adaptation is both productive and realistic. There is a level of paradox in the assumption and in the thought that it is only in light of adaptation that evolution makes sense. The fact is that adaptation obscures descent. Darwin saw the point, as Dov Ospovat makes clear in *The Development of Darwin's Theory* (1981). In the early decades of the nineteenth century, there was a broad recognition that animal form was to be understood in terms of the "conditions of existence" in which the animals lived. The thought had its roots in Cuvier and rapidly spread across Germany and to Britain. In Britain, the views found hospitable ground because they fit well with eighteenth- and nineteenth-century natural theology; but their appeal was broader than that, since German *Naturphilosophes* were equally taken with the idea. By the middle of the nineteenth century, many of the leading naturalists were challenging this teleological conception, including a wide range of figures from von Baer in Germany to Agassiz in the United States and Richard Owen in Britain. Darwin followed the dissenters after his return on the *Beagle*. The core idea to be confronted is that form follows function, that animals are exquisitely fit to their environment, and that this fit explains animal form. Here is Ospovat (1981, 26) on Darwin:

Darwin argued that the wings of flightless beetles cannot be explained by the teleological assumption that every organ is of some functional importance to its possessor. These wings show, he reasoned, that the beetles were "born from beetles with wings and modified—if simple creation, surely would have [been] born without them." It is not the mode of introduction, but the teleological assumption, that is at issue. If organisms are formed solely with reference to their conditions of existence, then heredity, and hence transmutation, are left out of the question.[15]

A host of issues is involved here. The first observation is straightforward: however complex wings might be, the presence of wings in flightless insects is not explained by their evolutionary function. The right explanation is that wingless beetles are descended from beetles with wings, and in which wings served their customary function. The more probing point is also more indirect, but it was equally clear to Darwin: in the case of flightless insects, it is the very fact that these wings do *not* function as wings that is most revealing of descent. It is most especially the nonfunctional aspects of animal form that reveal what Darwin later called "common descent."[16]

What has emerged in the twenty years since the "Spandrels" paper (though by no means solely as a result of that paper) is a much more eclectic and pluralistic research program in evolutionary biology, with more sensitivity to problems involving developmental limitations, more awareness of the deep conservatism in underlying genetic mechanisms, the significance of phylogenetic analysis and stochastic effects on evolution, and at the same time a more nuanced understanding of the action of natural selection. In some ways, we are left in a bit of a stand-off. It may be difficult to show empirically that reverse engineering is sufficient to identify adaptations. However, it is often difficult to show whether something is constrained in a way that limits adaptive response. These too cry out for evidence in their favor.

The critical methodological question for evolutionary biology, and for evolutionary psychology, is a practical one that is by no means simple: Can we, in practice, validate the evolutionary explanations offered for human psychology? What methods can we deploy that would make evolutionary explanations likely or unlikely in these cases? Ketelaar and Ellis think that the Lakatosian framework supports an affirmative stance for evolutionary psychology because it sets aside the problems of naive falsificationism. By assuming the "metatheoretical framework" of modern evolutionary theory and embracing the "middle-level theories" that apply it, such as Trivers's account of parental investment or Hamilton's work on kin selection, they claim that evolutionary psychologists can generate a variety of hypotheses that in turn can be confirmed. Indeed, they claim that these hypotheses are confirmed.

The focus of the current chapter is reverse engineering; other applications of evolutionary theory will occupy us in subsequent chapters. Ketelaar and Ellis defend a program of research that embraces, and depends on, reverse engineering. The question reverse engineering offers is one of how well an accepted model or theory explains some observed phenomenon. This fits a broadly Lakatosian framework, and the question should then be whether the resulting program is progressive or not. The methodology is directly relevant to the question of whether evolutionary psychology offers a progressive research program or a degenerative program that constructs epicycles to save itself from anomalies. It is not enough to find new empirical applications. Retrofitting of models to observations is not a difficult matter, so long as the constraints are relatively fluid. To use a well-worn example, the epicycles of

Ptolemaic astronomy allowed a very good fit to the data concerning apparent motion of the planets within a geocentric model of the universe. There was some variation in the number of minor epicycles (those around the planetary orbits) necessary to account for planetary motions, but it is clear that the resulting system was extremely accurate and flexible (see Kuhn 1957). It is likewise not enough to find some new predictions. The Ptolemaic system predicted eclipses reasonably well, as well as the irregularities in planetary motion known as "retrograde" motion. The Ptolemaic system nonetheless collapsed. What matters from a Lakatosian perspective is whether the predictions we have are based in empirically motivated models, with parameters, constraints, and design criteria that are independently established. The problem with epicycles is that they can be introduced for free, without independent empirical constraints.

It may be true that human reasoning can be understood in terms of the importance of social contracts and reciprocal altruism. It may be true that the development of daughters can be understood in terms of the importance of paternal involvement. It may be true that sexual selection is consistent with a sensitivity to features "indicative of good genes." Many of the suggestions Ketelaar and Ellis draw from writings on evolutionary psychology may be true. What is missing, and what is needed, is knowledge of a sort that contributes to a sound empirically motivated evolutionary model, including information concerning the sort of environmental "problem" cognitive mechanisms are responding to, the phenotypic and genotypic variation present, the structure of the relevant social groups, the gene flow between them, and other population parameters. Without such information, we do not know whether evolutionary psychology offers us epicycles or explanations.

7 True Causes

Given these different applications of engineering design, what can we conclude? And what can we say of these limited cases favoring evolutionary psychology? I think that, following the simplified fourfold scheme I've offered, we can begin to sort out the cases. I think this also begins to sort out the differences of opinion over the usefulness of reverse engineering, though I won't trace that issue out in any detail. The fundamental inferential task is to explain form in terms of function, beginning with the observed function. This *is* reverse engineering. We assume that there are *some* constraints on design. Even the pentadactyl hand shows structural similarities with the forelimbs of other vertebrates. What do we know of these constraints? How do we know about them? On the one hand, we may infer that the constraints on design derive from the materials at hand, or we may infer that there are physical constraints imposed by the environment. On the other hand, we may be able to specify these constraints *a priori* or only *a posteriori*.

Where we have *a priori* constraints, know what limits they impose on acceptable functions, and are offered principled reasons for recognizing either organismic or environmental limitations, the case for adaptation can be reasonably compelling. If, as is true, exoskeletons limit weight-bearing potential, it follows that there will be absolute limitations on size for arthropods. If, as is true, surface tension is fixed (given salinity), it follows that there will be limitations on organisms that can walk on water. These are constraints that shape evolutionary models, mirroring the way they shape evolution itself. Neither, I've suggested, is decisive, in the absence of historical information; but such cases do offer *substantial* support for adaptationist conclusions. So, to use the example I've leaned on at some length, *Archaeopteryx* is an intriguing case of design. The question is what the design might be *for*, what *Archaeopteryx* was designed *for*. Dennett contends that the *a priori* design considerations lead us to conclude that *Archaeopteryx* is designed for flight. I have argued, following some respected paleontologists, this is not so, and that it is more likely, on simple design considerations, though augmented by considerations of common descent, that *Archaeopteryx* was a terrestrial predator. Note that here the explanation is reinforced by considerations beyond those of simple design. In the case of *Halobates*, the case for design could again be compromised by historical information; it turns out that it is not. The cases offer us a methodological moral relevant to adaptationist reasoning: the a priori design considerations are inconclusive apart from historical information. The *a priori* constraints are defeasible in ways that are not trivial. In some cases, as we've seen, they are not only defeasible, they are defeated. Here we can defend the idea that we are not simply invoking epicycles, but true causes.

The alternative is to infer the constraints on design from the design itself, apart from any assumptions about historical function. These are the *a posteriori* cases of engineering design. When we have only *a posteriori* constraints on environmental structure, adaptation (or adaptiveness) is not tested as much as it is assumed. This is exactly what some defenders of optimality and adaptationism have insisted upon (e.g., Mayr 1983 or Maynard Smith 1978). If this is the goal, then there is no principled objection to the results, given the program. It is then no test of the "adaptationist programme"; no test of the claim that a specified trait is an adaptation; no test of the ubiquity of selection; no test of the idea that organisms are optimally adapted to their environment. It is but an application of the assumption. The problem with the program is that it says nothing about the history: It offers no conclusions concerning the genesis or the evolutionary history. It serves fundamentally to "save the phenomena," apart from any real concern over how they came to be. It neglects evolutionary questions. Finally, when we have only *a posteriori* constraints imposed after the fact on organismic structure, as in the case of the hand, again there is little substantial support for the adaptationist conclusions, apart from demonstrating consistency. This is where the agendas of Dennett, Buss, and Cosmides and Tooby have their proper home. They do not begin with the physical constraints, and they do not, intentionally, just *assume* design in order to *infer* form. They claim to have evidence concerning design based on form. Having assumed the "fact" of design, they "explain" the complex structures and behaviors we see as the consequence of natural selection, often without independent evidence. They do not argue *for* design but *from* form to function and then again *from* function *to* form.

The problem with this as a line of reasoning may be made clear by turning to an analogous case I've already alluded to, the theological use of reverse engineering. Hume showed in his *Enquiry Concerning Human Understanding* (1748) that the so-called argument from design is a failure, provided it is taken as a justification of faith from the "evidences" available to reason. Somewhat paradoxically, Hume died before Paley mounted his famous version of the argument from design. The classic refutation thus predates the best statement of the case. In section 11 of his *Enquiry,* Hume contends that design could offer no reasonable ground to support the Christian religion. The key problem he identifies is this:

You find a certain phenomenon in nature. You seek a cause or author. You imagine that you have found him. You afterwards become so enamoured of this offspring of your brain, that you imagine it impossible, but he must produce something greater and more perfect than the present scene of things, which are so full of ill and disorder. (Hume 1777, 11.14)

The problem, notice, is not the lack of effects. Hume allows that we may infer causes from observed effects. He insists, though, that when we do so "we must proportion the one to the other, and can never be allowed to ascribe to the cause any qualities, but what are exactly sufficient to produce the effect" (ibid., 11.12). The problem Hume observed among natural theologians was that they were not content with causes proportioned to the observed effects. Instead, they drew grand, and unwarranted, conclusions concerning the world and the attributes of their God. This application of reverse engineering has the same structure as that advanced when we have only *a posteriori* constraints on form and infer an adaptive cause. It is the same argument in a different context.

It is not unreasonable to insist on a more restrictive standard, given that there can often be more than one cause for any observed effect. The key point was made elegantly by M. J. S. Hodge, in conjunction with the idea that Darwin required a *vera causa* ideal for a scientific theory.[17] The roots of the ideal lie in Newtonian physics, and how it was understood in the mid-nineteenth century by people such as Whewell. In the nineteenth century, the idea of a *vera causa*, or a "true cause," required independent confirmation. Hodge (1990, 236) explains:

So the *vera causa* ideal, as Darwin sought to conform to it, required that any cause introduced in a scientific theory should be not merely adequate to produce the facts it is to explain on the supposition that it exists. For the existence of the cause is not to be accepted on the grounds of this adequacy. Its existence should be known from direct independent evidence, from observational acquaintance with its active presence in nature, and so from facts other than those it is to explain.

There are three required pieces of evidence of something's being a true cause: first, it is necessary to provide independent evidence concerning the existence of the cause; second, it is necessary to show the cause is sufficient to explain what was observed; and third, it has to be responsible for the specific facts needing to be explained. Whereas Hume insisted that we could not attribute to the cause anything more than is necessary for the observed effects, Whewell insists that without independent evidence concerning the cause, we should not even do that much.

The most suspect cases of reverse engineering fail not only the more restrictive limits imposed by the ideal of Whewell's *vera causa*, but even the more modest limitation imposed by Hume.[18] The more compelling uses of design, such as the foot of *Halobates* or the design of prairie dog burrows, do offer us independent evidence concerning the causes. That is precisely what makes them appealing. Dennett, by contrast, feels no need to provide independent evidence relevant to the design of the foot of *Archaeopteryx*; and I've even argued that when we look at the independent evidence, the case collapses. If we consider the third condition, that the hypothetical "cause" must be responsible for the observed effects, even the case of the hand is problematic, mostly because it is underdescribed. Griffiths (1996, 526) observes, with a bit of hyperbole: "A trait like the pentadactyl limb of tetrapods, however, seems to have an enormous amount of inertia. The relative positions of its parts are preserved in everything from a frog's leg to a bat's wing. This seems to be a 'Newtonian' trait." There seems to be significant inertia to pentadactyl design, at least within the vertebrate lineage. Of course, *Archaeopteryx* and birds are exceptions that show that the trend can be modified. Though the hand is doubtless adapted for grasping, it is an instance of the pentadactyl limb; and there are constraints that are much broader and deeper than any simple inference from the current function to its form would indicate.

When we turn to evolutionary psychology and its use of reverse engineering, we find no attempt to provide independent evidence concerning the causes. There is no attempt to show that the hypothesized cause would be sufficient. In Buss's exposition of the case for jealousy, there is no significant appeal to the historical conditions of human evolution, aside from general appeals to the conditions of the Pleistocene. He is content to offer evidence from social psychology. Even if this is good psychological evidence concerning current differences between the sexes, it does not give us more than that. We can, of course, find some phenomena—perhaps the patterns of sexual jealousy are one of them—which can be brought into conformity with some evolutionary models. Without knowing, however, the conditions in which jealousy evolved, it is impossible to know whether we have the right set of causes or the right explanation. Ketelaar and Ellis would turn the methodology into a mark of virtue. Assuming that the causes are known, in at least general terms, they essentially turn the investigative process into one of finding the particular constraints relevant to the case.

The evolutionary psychology cases do not even meet Hume's standard. This is what we find in the most aggressive forms of reverse engineering. When we proportion independently verifiable causes to explain the effects we observe, as in *Halobates*, we are fully in conformance with Hume's demands. The causes are proportional to the effects. We even have the independent evidence that marks a true cause. When we reverse the direction and infer constraints to explain what we observe, then we invoke hypothetical causes. As Hume (1777, 11.16) says, when we infer causes from effects, and then reconceive the causes in terms of the effects we infer, we "have aided the ascent of reason by the wings of imagination; otherwise they could not thus change their manner of inference and argue from causes to effects."

3 The Dynamics of Adaptation

1 Human Language and Cognition as Adaptations

Human language and cognition are natural candidates for inclusion within an evolutionary psychology, where this is thought of as a theory of evolved psychological mechanisms. As I've observed, advocates of evolutionary psychology insist that features with complex functional designs must somehow have evolved, and must have done so because they provided a substantial advantage to our ancestors. Their presence, that is, must be due to natural selection (Tooby and Cosmides 1992, 49ff.; Pinker and Bloom 1990). The classic example is the vertebrate eye, with a carefully orchestrated structure designed to focus light and transmit information. That was William Paley's most famous example of design in the early nineteenth century, though not his only example. As Darwin (1895, 186ff.) recognized, such "organs of extreme perfection and complication" must be explained as the consequence of natural selection acting on "numerous, successive, slight modifications" over generations. They cannot be the consequence of chance, but must somehow result from "design," although as Darwin saw, they are the result of natural design rather than that of a deity.

Human language and human reasoning are certainly complex capacities, and they are *evolved* capacities, which means they were not present, or not present to the same degree, in our ancestors. Given their complexity, we might also expect them to be, directly or indirectly, the consequences of selection. However, a discussion of what we actually *know* about the evolution of either human language or cognition would be very brief. The same would not be true of, say, a discussion of the eye. Evidence concerning either language or cognition, as we will see, is sparse even if suggestive. Speculation concerning the evolution of both language and cognition is nonetheless very common. To take a philosophical exemplar, Robert Nozick was a famous twentieth-century philosopher who made foundational contributions to the understanding of human understanding and the logic of inference. He contends that accounts of human rationality can resolve problems of (nonenumerative) induction only by appeal to rules or principles of reasoning, and suggests that the source of these rules may lie in our evolutionary history. Rationality, he thinks, depends on reasons reliably produced in accordance with such rules, and natural selection should favor capacities whose exercise has increased fitness as a consequence of their reliability (Nozick 1993, 113). So we form expectations based on experience. These are inductive inferences. What makes them *inductive* is essentially that they are open ended.

Perhaps a slightly different example will help to illuminate the thought. Mendel's peas were either wrinkled or round. He knew the ratio of round to wrinkled seeds in the first generation of offspring because he counted them. Of 7,324 seeds in the first generation, there were 5,474 round ones and 1,850 wrinkled ones. The ratio he reports is 2.96 to 1 (Mendel 1865). That is not induction; it is just counting and calculating. Mendel *inferred* or *induced* that the ratio of dominant to recessive traits in the first generation of hybrid offspring should be 3 to 1. That is properly a case of induction, and not simply counting. Nozick, like many philosophers, would like to discover rules of induction, determining which rules make such inductive inferences reasonable or rational, and, by exclusion, what makes other inductive inferences unreasonable or irrational. It would be reasonable to think of a theory of inductive reasoning, so understood, as a theory concerning the principles responsible for rational expectation. This is precisely how Nozick thought of the issue.

Some of our inductive inferences are very simple, such as our expectation that the sun will warm us or that water will quench our thirst. Some are more complex, such as our predictions concerning the behavior of those we know, based on their character. Some, of course, are still more complex, such as Mendel's inferences concerning hereditary patterns among plants. Some of these expectations, or predictions, or inferences, are reliable. Some are not. Some are rational; some are not. Nozick's key thought is that what distinguishes rational inferences from those that are not is the rules or principles that guide them. These rules may be explicit, or they may be assumed without reflection. Whatever these rules might be, they need not be the result of reflection on experience. Some, of course, are. We learn, painfully, that we can be burned by the sun as much as we can be warmed by it. We learn that smiles can deceive as well as inform. We are taught that it is irrational to ignore base rates. These are things learned more or less explicitly, if at all. But not all reasoning could be based on such learned rules. Some such rules *must* be assumed; otherwise, we would have no rules to start with. We might nonetheless *discover* what these assumed rules are. That is the business of psychologists and of empirically minded philosophers. Among other things, we uncover the rules that guide human reasoning and how we come to them. If Nozick is right and those rules have evolutionary roots, then it is the business of evolutionary psychologists to uncover them.

Why do we need rules of reasoning at all? What is it to be guided by rules of reasoning? Of course, it need not mean that we can rehearse the rules, or easily make them explicit. I think this is the thought that guides Nozick and many others: in a locally stable and predictable environment, simple reflexes might suffice to guide our choices.[1] Humans, however, are animals that respond to wide and unpredictable variations in circumstance. We had to. We are large animals with large ranges. Larger ranges generally mean variation in circumstance. Likewise, we are relatively long lived, and that too makes variation in circumstance more likely. Fruit flies, by contrast, don't experience even seasonal changes, though they are adapted to them. Their lives are just too short. We do experience seasonal changes and need to modify our behavior accordingly. For humans, grasping changes over geological time is a stretch, which may be part of why people get confused over global warming. In any case, our adaptation to longer time spans, with the associated temporal and spatial variability, Nozick thinks, is the key to understanding the evolutionary function of rationality. He says:

Rationality may have the evolutionary function of enabling organisms to better cope with new and changing current situations or future ones that are presaged in some, possibly complex, current indications. (Nozick 1993, 120)

Nozick is interested, in part, in articulating this "evolutionary function" of rationality, which, he agrees, depends on the processes that historically shaped and maintained it. This would make reasoning an adaptation, in evolutionary terms. He is mostly interested in what the rules might be. It is one question to wonder what rules might be most effective in guiding us now; it is another to ask what rules did guide us. The latter alone defines the "evolutionary function" of those rules. That is a concern Nozick shares with evolutionary psychologists. Nozick is studiously vague about precisely what this function might be, beyond the idea that rationality serves to enable us to cope with "new and changing situations."[2]

Evolutionary psychologists typically eschew general-purpose mechanisms in favor of cognitive mechanisms that are content specific and specialized. Nozick, as I have said, is a friend of general-purpose rules.[3] Evolutionary psychologists will, of course, grant the point that humans adapt to a wide variety of circumstances. In place of the more general rules that Nozick favors, evolutionary psychologists lean toward rules with more limited scope. Leda Cosmides and John Tooby, for example, suggest that human reasoning consists of a set of mechanisms organized around social exchange. The empirical studies of human reasoning by Cosmides and Tooby are some of the most elegant pieces of *psychological* work within evolutionary psychology. They have made substantial and important contributions to the study of human reasoning. I will discuss them at a bit more length later. They say, at one point,

> According to the evolutionary psychological approach to social cognition . . . the mind should contain organized systems of inference that are specialized for solving various families of problem, such as social exchange, threat, coalitional relations and mate choice. . . . Each cognitive specialization is expected to contain design features targeted to mesh with the recurrent structure of its characteristic problem type, as encountered under Pleistocene conditions. Consequently, one expects cognitive adaptations specialized for reasoning about social exchange to have some design features that are particular and appropriate for social exchange, but that are not activated by or applied to other content domains. (Cosmides and Tooby 1992, 166)

The evolutionary function of human reasoning would then be to facilitate and monitor social exchange. Other applications of human reasoning that are, though complex, manifestly not adaptations—playing chess, programming computers, and the like—can then be treated as spandrels. There are, of course, many other cases to which we can and do apply our cognitive capacities, from astrophysics and evolutionary biology to cooking and carpentry. The *evolutionary* function of human reasoning, however, is to facilitate social exchange, if Cosmides and Tooby are right. At least this is the evolutionary function behind some of our cognitive adaptations.

Typically, this emphasis on domain-specific mechanisms, which is more prevalent within evolutionary psychology, is connected with what is called *modularity*. Jerry Fodor (1983) provided a classic account of modularity within psychology. What matters here is mostly one condition he emphasized, *encapsulation*. Roughly, the idea is that, for example, the color we see does not depend on what we expect, the sound we hear does not depend on our beliefs about the state of the economy, and the word we hear is even remarkably independent of the specific sounds we encounter. I am broadly sympathetic with modularity, but I don't wish to take on those issues here. I'm content to notice the variety of opinion.[4] My focus here is not on such questions of what is called the "architecture" of the mind. Although it might be interesting to pursue the question of the evolvability of modular as opposed to nonmodular organization, I won't. I think that absolutely nothing I say depends on this issue.

In a parallel vein, Steven Pinker says the conclusion is "inescapable" that linguistic capability is the consequence of a history of natural selection. Speaking of the "language instinct," he says:

Every discussion in this book has underscored the adaptive complexity of the language instinct. It is composed of many parts: syntax, with its discrete combinatorial system building phrase structures; morphology, a second combinatorial system building words; a capacious lexicon; a revamped vocal tract; phonological rules and structures; speech perception; parsing algorithms; learning algorithms. Those parts are physically realized as intricately structured neural circuits, laid down by a cascade of precisely timed genetic events. What these circuits make possible is an extraordinary gift: the ability to dispatch an infinite number of precisely structured thoughts from head to head by modulating exhaled breath. (Pinker 1994, 362)

The evolutionary function of this complex apparatus, Pinker thinks, is communication by speech. It is clear enough that this is not the only thing we do with language. We can communicate, but we can also talk to ourselves, and deliberate. We can rant and rave. We can write poetry, and we can simply play with words. We can write books. The evolutionary function that explains the prevalence of language in contemporary human populations, though, is supposed to be verbal communication. Writing poetry or books is at best a happy side effect. Ranting is also a side effect, though perhaps not a happy one.

There are some who have a much more negative assessment of the role of natural selection in explaining such cognitive capacities. Some are experts on human language. Noam Chomsky (1972, 70), probably the single most influential figure in the history of thought about human language, is striking for his dissent:

As far as we know, possession of human language is associated with a specific type of mental organization, not simply a higher degree of intelligence. There seems to be no substance to the view that human language is simply a more complex instance of something to be found elsewhere in the animal world. This poses a problem for the biologist, since, if true, it is an example of true "emergence"—the appearance of a qualitatively different phenomenon at a specific stage of complexity of organization.

I am not actually much troubled by the prospect of emergence, understood this minimalist way, and neither is Chomsky. It would be easy to overinterpret Chomsky. I think many do. As far as I can tell, Chomsky does *not* deny that language confers a substantial advantage, or that therefore it could be subject to selection. That would be terribly silly, and he is not prone to silliness. He certainly does not deny that human language evolved. That would be even sillier. However, he sees human language as so unique, so different from other communication systems, that it provides a case of the "emergence" of a qualitatively distinct phenomenon. It is interesting, and mildly paradoxical, that the very features that lead Chomsky to evolutionary despair are the very same features that convince Pinker that an adaptationist explanation of linguistic capacities is unavoidable. I will argue in what follows that we do not have anything approaching an *explanation* of the unique features of human language in evolutionary terms. We have no substantial, empirically motivated, explanation of the evolutionary function of language. I do not deny that there is an evolutionary history, and I do not deny that this history shaped the languages we have; I do deny that the capacity to use language made us what we are as creatures, in many ways. I will contend, though, that the sort of evidence we have does not support the stories we tell about ourselves. They may be comforting as stories, but they do not have the sort of evidence we should require of them. We might as well explain the structure of orchids in terms of their beauty: it may be comforting, but it is not scientific. My moral is much closer to Chomsky's than Pinker's. It is, as far as I can tell, indistinguishable from Chomsky's.

Richard Lewontin is equally skeptical about the evolution of human cognition. Again, Lewontin incontestably believes human cognition evolved. He likewise knows that human language evolved. To believe otherwise would be errant nonsense. He does say this concerning the evolution of cognition:

Despite the fact that there is a vast and highly developed mathematical theory of evolutionary processes in general, despite the abundance of knowledge about living and fossil primates, despite the intimate knowledge that we have of our own species' physiology, morphology, psychology, and social organization, we know essentially nothing about the evolution of our cognitive capabilities and there is a strong possibility that we will never know much about it. (Lewontin 1990, 229)

Lewontin is obviously *not* skeptical about evolution, or even about the evolution of human cognitive capacities. He is, after all, a distinguished evolutionary biologist. His very business is the construction and validation of evolutionary explanations. His principal business is fruit flies, and he knows a great deal about their evolution. He also knows a great deal about evolution in general. He *is* skeptical about the theories invented to explain human cognition. I largely follow his lead.

My own attitude toward the claims of evolutionary psychology is skeptical in the same way, and I underwrite the skeptical moral here by turning to a different class of evolutionary models. As I've said at the outset, my skepticism is not grounded in any skepticism about evolution in general, or about the importance of natural selection. Evolution is no longer a matter of rational dispute, and neither is the existence or importance of adaptation through natural selection. Natural selection is not the only factor relevant to evolution, though it is an important one for understanding biological form and function. My skepticism is also not meant to reflect any skepticism about human evolution in particular, or to underestimate the considerable knowledge we have of ancestral hominids. I do not doubt that human language and human rationality are evolved capacities, or even that they provided substantive advantage to our forebears. Perhaps they are adaptations; I do not claim to know. Indeed, I claim *not* to know. I also do not doubt that a great deal is known about evolutionary history that is potentially relevant to speculations about our evolutionary history. But I am skeptical about the explanations offered. There is a dilemma in these cases. On one side, the problem is that the actual explanations are sketchy and indeterminate. This leaves them unable to engage the features they are supposed to explain. The other horn of the dilemma is that if they were more specific, and more determinate, they are then not supported by the available evidence. Thus they become either irrelevant or unsupported, and sometimes both.

I will urge, in particular, that the cases of human language and cognition are inhospitable to systematic evolutionary analysis. This is particularly true when the aim is to explain human language and cognition *as* adaptations. These difficulties derive fundamentally from the fact that the claim that these capacities are adaptations is a historical claim, together with the absence of the right kind of evidence to address the historical issues. As G. C. Williams (1966, 5) emphasized in his classic *Adaptation and Natural Selection*, "adaptation is a special and onerous concept that should be used only where it is really necessary." Reconstructing an evolutionary history or the factors that shaped it is a complex, difficult problem of inference. Discoveries are made almost daily that can reconfigure our assessment of evolutionary history. In the case of human evolution, one problem is that there are no suitably close living relatives. We diverged from our closest living relatives among the apes perhaps seven to ten million years ago; some recent genetic estimates suggest only six million years ago, but the fossil evidence is pretty unambiguous. There is a relatively rich fossil fauna, however, covering the last four million years or so. Among these, there are at least three species within our own genus, *Homo*, including *Homo habilis*, originating roughly 2.4–1.6 mya, *Homo erectus*, dating from 1.8 mya, and *Homo sapiens*, dating from as much as 400,000 years ago, but most likely much less than that. For all the richness of the work that has been done with the fossil finds, and in spite of the considerable insight this work has yielded into human evolution, little of it bears directly on the questions of particular interest to evolutionary psychologists. The record is certainly sufficient to support the idea that humans are the product of a long history of evolution; but it does not offer enough to support anything that would suffice as an explanation of the process, much less an explanation of human language and reasoning as adaptations. We will return to the specifics in the next chapter.

The speculations we are offered under the rubric of evolutionary psychology are not substantive enough to count as anything more than crude speculations. The project in evolutionary psychology, as I've already said, is not simply one of showing that humans evolved, or that human psychological characteristics are adaptive. Evolutionary psychology claims to give an explanation of the emergence of particular traits in terms of natural selection. Insofar as what is offered purports to be an *explanation* of human language or cognition in terms of natural selection, evolutionary psychology falls well short of the mark.

This chapter takes on one more piece of the task. Population genetics gives us one straightforward protocol to evaluate evolutionary explanations in terms of natural selection. This is what I've called "adaptive thinking," reasoning from evolutionary challenges to adaptive responses. Buller (2005, 69) calls this "evolutionary functional analysis." So, to use his example: human males compete for females; adaptive thinking requires determining what strategies or behaviors would facilitate male access to females. There are many uncertainties in applying the method, and it is often applied improperly. However, it offers one way of understanding adaptations; it gives us one way to see how such explanations are confirmed; and it incorporates a set of conditions for explaining adaptations. Unlike reverse engineering, this is explicitly a historical method; accordingly, it is prospective, or forward looking. It insists that adaptation explanations are historical explanations. It constructs historical explanations empirically. So knowing whether some explanation is a good one depends on knowing what historical conditions did, or did not, obtain. It is also explicitly an empirical method. It insists that *knowing* whether an explanation is adequate depends on *knowing* whether various conditions apply. If there is uncertainty concerning the conditions, then there is uncertainty concerning the conclusions. If the conditions are unknown, then the conclusion is likewise unknown.

I think of the explanatory models as dynamical models. Population genetics specifies a set of conditions that are sufficient to guarantee that there will be evolution by natural selection; indeed, a model that specifies the values of the various parameters will tell us what changes would be expected. It will thus be a *predictive* model: given any set of values for the parameters, a model will specify some evolutionary outcome, or some range of evolutionary outcomes. In terms of evaluating the theories, then, we need to reverse the inference. This is a properly *epistemological* problem, concerned with whether we *know* a claim is true or with *how* we know a claim is true. The observed empirical data will always be less clear than we would like. We begin with data that is ambiguous; we need to infer what happened from what we now see. The epistemological problem is to determine which causes yielded the results or outcomes we see. This will depend on knowing what alternative theories or models we have to explain the process. If we have several that are consistent with what we see, we should then ask which is more likely. We should also ask, of any proferred explanation, whether it has empirical data of a sort that it requires. This again is a comparative question. That is, we see some outcomes. We are offered an explanation, or more than one explanation. If there is more than one possible explanation, the epistemological problem is whether there is enough evidence to favor one explanation over another; if there is only one explanation, the epistemological problem is whether there is enough evidence to embrace it. In some cases, this is a straightforward empirical matter. In other cases, it is not. So, to revert to an example I've already used, geocentric and heliocentric models were both able to explain the astronomical observations available in the seventeenth and eighteenth centuries. The reasons favoring heliocentric models were not simply empirical ones. There were, of course, empirical phenomena needing to be explained: observed planetary motions. No theory would be acceptable that did not explain retrograde motion, for example. The evolutionary question is also finally a comparative one, but one with a specific focus. I want to ask whether the sorts of explanations offered within evolutionary psychology are adequate given the constraints there are on otherwise well-established evolutionary and adaptive explanations. Population genetics thus offers one paradigm for evaluating explanations in terms of adaptation.

2   Adaptation and Adaptationism

To see why I think the discouraging moral is also the reasonable conclusion concerning evolutionary psychology, we need to focus on a prior question: What is necessary in order for a trait to be an adaptation? Correspondingly, the epistemological questions are these: What do we need to know in order to know that a trait is an adaptation? Do we know these facts? Are we likely to know them? The questions obviously are connected. An *adaptation* is a trait that is present, or was maintained, because of the selective advantage it offered to ancestors; in this sense, to claim that something is an adaptation is to make a historical claim (see, e.g., Lewontin 1977, 1978; Brandon 1978, 1990; Burian 1983; Sober 1984; and Griffiths 1996). Traits may be present because of chance, or drift; they may be developmentally or genetically linked to traits that are selected for; or they may be genetic side effects. In any of these cases, a trait may be beneficial and improve fitness, but that alone would not make it an adaptation. To repeat a common theme of Stephen Gould's, a trait may be the result of causes other than those that make it useful. A trait is an adaptation only if its presence is due to the fact that it conferred greater fitness on previous generations. Whether a trait is an adaptation thus depends on its evolutionary history; and explaining some trait as an adaptation depends on knowing the evolutionary history that produced it. An ideal adaptive explanation also needs to reveal what an adaptation is an adaptation for.

This is the root problem that evolutionary psychologists face: they claim that human psychological traits are adaptations and thus need to *explain* them *as* adaptations. In an ideal world, we would look to lineal ancestors to gain the information we need concerning the kind and extent of variations, their relative adaptedness, the ecological pressures that influenced them, and their heritability. This is rarely possible, and we are forced to look to related species or groups in order to get some sense of the kind of factors that are relevant to understanding evolutionary history. The explanatory problem lies at one remove from the historical problem, but it inherits the problems it poses: if a trait is an adaptation only because its presence is due to the fact that it enhanced fitness, then showing that a trait is an adaptation requires showing that its presence is a result of the enhanced fitness of ancestors who had that trait. If selection depends on population structure and heritability, then showing that a trait is an adaptation requires evidence concerning population structure and heritability. This is the historical problem. As the case of *Archaeopteryx* illustrates, sometimes the indirect evidence can be compelling. The explanatory task is then one of reconstructing evolutionary history from a contemporary record, together with some modest historical information. So we look to the structure of the foot, and feathers, and to relatives. We can often piece together a defensible explanation. This is not an impossible task, but it is a difficult one.

If a defensible dynamic explanation can be constructed for psychological capacities, it would answer perfectly to the ambitions of an evolutionary psychology. Here again is a representative passage from Cosmides and Tooby (1994, 534):

Like a key in a lock, the functional organization of each cognitive adaptation should match the evolutionarily recurrent structural features of its particular problem domain. . . . Because the enduring structure of ancestral environments *caused* the design of psychological adaptations, the careful empirical investigation of the structure of environments from a perspective that focuses on adaptive problems and outcomes can provide powerful guidance in the exploration of our cognitive mechanisms.

In order to reasonably embrace an evolutionary explanation, we at least need empirical evidence concerning the structure of the ancestral environment and the adaptive responses of our ancestors in those settings. If we want to know what the lock is, we need to know what the "evolutionarily recurrent structural features of its particular problem domain" actually were. The question, more specifically, is what kind of information we need in order to support a reliable evolutionary explanation for an adaptation. In *Adaptation and Environment* (1990), Robert Brandon lays out five conditions on what he calls "adaptation explanations." As he describes the five conditions, they are qualitative. Serious evolutionary studies deploy quantitative standards. This makes the actual evolutionary requirements both more reasonable and more demanding. Nevertheless, these five conditions provide, in Brandon's view, something like an ideal type sufficient for explaining adaptations, in the sense that a full and complete evolutionary explanation of an adaptation would draw on evidence to answer each of the conditions, though he admits we often have cases in which we have only partial answers. Perhaps most evolutionary explanations even fall short of Brandon's ideal in one way or another, but the five conditions nonetheless provide a reasonable yardstick to consider in assessing the qualitative adequacy of specific adaptive explanations. The five conditions offer no simple litmus test for good adaptive explanations, but they do capture the sort of criteria relevant to deciding whether an explanation passes muster. They reflect Lewontin's simple rule that evolution by natural selection requires *heritable variation in fitness*. These are the five conditions for qualitatively sufficient explanations of adaptation based on natural selection (Brandon 1990, chap. 5):

(1) *Selection*   There must be evidence that selection has in fact occurred: we need to be able to distinguish the possibility of selection from other candidate evolutionary explanations and to demonstrate something about the character and strength of selection, given that it is a factor. It is sometimes possible to do this directly, by measuring variations in fitness to measure selection (see, e.g., Lande and Arnold 1983), though this is often difficult. It is particularly difficult in the case of behavioral traits (see Arnold 2002). This will require, among other things, that we have some information about the character and extent of variation in the ancestral forms, since without variation there can be no selection. It will also require information about the differential survivorship and reproduction of those various ancestral forms (see Kingsolver et al. 2001). Selection is, after all, simply the differential survival and reproduction of one alternative in comparison to others.

(2) *Ecological factors*   There must be an ecologically based explanation for the selection observed: we need to mark off some ecological factors that offer an explanation for the presence and the strength of selection, whether these are factors in the external environment or in the social environment. The range of such factors can be considerable. There may be abiotic factors that limit

survivorship, such as the effect of severe winters on the survivorship of birds. There may be biotic factors that also affect relative survivorship, such as the impact of predation or competition for resources. These will sometimes involve competition for mates rather than survivorship. All of these could have selective effects. A good explanation would tell us specifically what factors determine what selection takes place.

(3) *Heritability* Differences among individuals must be heritable: this means that there must be a correlation between the phenotypic traits of parents and offspring that is greater than would be expected by chance; that is, there must be a correlation between parents and offspring that is greater than the correlation between arbitrarily chosen individuals. This essentially requires that whatever differences affect selection among parents should have similar, though perhaps not identical, expression in the offspring. The differences that matter in one generation need to be passed on to the next generation if they are to have an evolutionary impact. It is sometimes possible to show that a trait runs in families, and that too can be used in evolutionary arguments in lieu of a formal estimate of heritability (see Arnold 1994). The importance of estimating heritability is that this allows us to see that selective differences are passed on from one generation to the next.

As a bit of an aside, here is what seems to be a remarkable mistake from Cosmides and Tooby, which I am not able to fully explain. It is hardly characteristic. It is nonetheless an important issue. They say this:

Contrary to popular belief, developmental evidence is not criterial [for an adaptation]: Adaptations need not be present from birth (e.g., breasts), they need not develop in the absence of learning or experience . . . and they need not be heritable. (Tooby and Cosmides 1990, 380)

In another context, they say something similar:

In fact, although the developmental processes that create adaptations are inherited, adaptations will usually exhibit low heritability. Differences between individuals will not be due to differences in their genes because adaptations are, in most cases, universal and species-typical. (Cosmides and Tooby 1992, 180)

This is almost a primer for mistakes concerning the analysis of heritability. For the moment, let's set aside the question of the relevance of development. It is at least true that adaptations need not be present from birth. Any secondary sexual characteristics—beards in some human groups, or red color in male cardinals—would suffice to make that point. I want to focus at this point on heritability.

Heritability has a precise definition within evolutionary biology. In what is often called the "broad sense," heritability is the proportion of the total variance that is due to genetic variance. Consider a simplified case. Suppose we have a population of plants with three different genotypes, two homozygotes, and one heterozygote. If we look at each genotype, there likely will be variations in height among individuals. If we graphed the heights of each group, we would find an array of values. The variance is simply the scatter of heights across the range. If they are well behaved, these will be normal distributions, the bell curves statisticians love. We could also plot the heights of all individuals. The scatter among individuals here is the total variance. The genetic variance is, by contrast, the spread of the mean values for the several genotypes. If the mean values for the genotypes do not differ very much by comparison with the total variance, then the heritability will be low. Most of the variance is then environmental variance. If the mean values for the genotypes differ a good deal by comparison with the total variance, then the heritability is high.

In what is called the "narrow sense," heritability is limited to *additive* genetic variance. It is the proportion of total variance that is due to additive genetic variance. This turns out to be very important for evolutionary explanations since it provides the response to selection. Initially, we can focus on the broad sense of "heritability." Exactly the same issue can be raised in either case. Notice what Cosmides and Tooby are saying. First, they say that adaptations will usually exhibit low heritability. That is simply not true. Many adaptations are heritable. Among humans, tolerance for milk products—lactose tolerance—is an adaptation. It also exhibits high heritability. Among humans, hemoglobin types are sometimes adaptations. Resistance to malaria is a consequence of sickle cell genes in a heterozygous state. It is perhaps the most studied case of adaptation within humans. It is heritable. Heritability is simply a measure of patterns of relationship and inheritance among individuals. Second, Cosmides and Tooby say the heritability of adaptations will be "low" because they are universal. Again, we need to focus on the careful formulation typically used by evolutionary biologists. If there is no difference within a population, then the heritability is actually undefined, because the variance in the population is zero. Even if the variation is "low"—and the variance is therefore not zero, but minimal—that says nothing whatsoever about the heritability of the variation. Again, heritability is the *proportion* of the total variance due to genetic variance. I don't think it makes much sense to think of "high" or "low" variance in absolute terms. Similarly, "tall" and "short" have meaning only within a population, by comparison with some group or

Box 3.1
Heritability and the analysis of variance

The variance for some value such as height tells us how close a randomly selected individual would be expected to be to the average for the population. This is calculated by taking the sum of the squares of the differences of observed values and subtracting it from the arithmetic mean within a population. So if x is the mean value for some set of values $\{x_1 \ldots x_n\}$ that occur with frequencies $\{f_1 \ldots f_n\}$, then the variance V for the population is simply:

$$V = \sum f_1 (x_n - x)^2.$$

Alternatively,

$$V = \sum f_1 (x_n)^2 - f_1 x^2.$$

Squaring the differences is an analytical tool to ensure they will always be positive. Also, a value further from the mean will contribute more to the variance. So the variance value is more sensitive to scatter from the mean than is a simple difference measure. Nonetheless, variance can be thought of as a measure of the scatter around a mean value. With a low variance, individuals would be expected to be nearer the mean, and with a high variance, they would likely be further away.

Francis Galton thought of *heritability* as the ratio of the deviation of offspring averages and midparent values from the population mean. This is broad heritability. So if x is the average value within a population, X_o is the average offspring value, and X_p is the average parental value, then the Heritability H^2 is the ratio of the differences of the parental from the populational averages. That is:

$$H^2 = (X_o - x)/(X_p - x).$$

What Galton found is that variance diminishes over generations, so that this ratio described what we call *regression to the mean*. Intuitively, a high heritability for some trait means that parental values are good predictors of that trait in offspring. What Galton was interested in was not whether parental values were predictive of offspring values, but the extent to which this is so. So with a heritability of 1.0, a trait in the parents is a perfect predictor of the trait in the offspring, and with a heritability of 0.0 the trait in the parents is irrelevant to the trait in the offspring. Galton calculated a heritability of 0.65 for height, which is a relatively high value for heritability.

For many purposes, the differences do not matter, but we now treat the heritability of a trait as the proportion of the phenotypic variance due to genetic variance. So in the broad sense, if we now take the genetic variance V_g, to be the variance for some phenotypic value among the various genotypes, and the total variance V to be the variance in the population, then heritability, again in the broad sense, is the ratio of the two:

$$H^2 = V_g/V.$$

This is not quite Galton's measure, but it approximates it. With a similar environment across generations, Galton's measure is no different from those more common in population genetics for broad-sense heritability.

Heritability, so defined, depends crucially on environmental conditions. The total variance depends on both the genetic and environmental components of variance. That is in fact the definition of total variance. So as the environmental variance is reduced, the total variance approaches the value of the genetic variance, and H^2 approaches 1.0. Similarly, as the genetic variance is reduced (say, through inbreeding, or from directional selection), the total variance approaches the environmental variance, and H^2 approaches 0.0. Heritability may thus be increased by reducing the environmental variability, and it may be reduced by reducing the genetic variability. Since heritability is simply the ratio of the variances, anything that increases the ratio increases heritability. If the environmental

Box 3.1
(continued)

conditions change, then so can the heritability. For example, a change in nutritional regime can change the heritability. It is enough if there is a reduction in the environmental conditions affecting variation. We sometimes find that improving a nutritional profile reduces the variance between groups. Alternatively, we may simply shift nutritional regimes to make them less different. That means that what is called "between-group" variance is reduced. This amounts to reducing the environmental component; if this reduces the total variance, it thereby increases the heritability within the group. Although it might seem paradoxical, the within-group variances may remain unchanged.

Estimates of what is called *additive genetic variance*, and therefore of heritability in the "narrow" sense, are often difficult to obtain. Experimentally, they depend on relatively large populations. It is this narrow sense of heritability that is crucially relevant to studies of selection. What is called the *response to selection* is a function of both additive genetic variance and the strength of selection. A higher correlation between phenotype and fitness (i.e., selection) affects the rate of evolution proportionally to the heritability. This result is foundational for studies of selection in the wild (see, e.g., Lande 1976). In field studies, the observed variance typically underestimates the variance in a population, and so we correct for that error, treating the observed values as a random selection from the actual values.

standard. Given this caveat, whether variance is "high" or "low" is simply not the issue. The genetic variance cannot exceed the total variance, of course, so the genetic variance will be low if the total variance is low.[5] The total variance may be small and all due to genetic variance. In that case, heritability is perfect; that is, it takes the value of one. The total variance may be large and not at all due to genetic variance. In that case, the heritability is zero. The amount of variance does not determine what proportion of it is due to genetic variance, that is, what is due to *heritability*.

What Cosmides and Tooby say is simply incoherent if heritability is understood in the way population geneticists use it. Perhaps all Cosmides and Tooby intend to deny is that in contemporary populations, heritability may be low because of selection in the past. It is at least true that with strong directional selection, additive genetic variance can be used up; but then under strong directional selection, as they surmise, variance may be reduced over time. However, that will be true when the variance is largely additive genetic variance, and so, when the heritability is high. With low heritability, selection will have little effect. So I am at a loss to explain why Cosmides and Tooby reject the importance of heritability. It is, in any case, integral to evolutionary models that the variance present in ancestral populations be heritable. Without heritability, selection is impotent.[6]

(4) *Population structure*   There must be information concerning the environment, population structure, and gene flow: the effective size of populations,

population structure, gene flow and interbreeding, as well as mutation rates, all affect the rate of evolution; we need to be able to fix these parameters if we are to distinguish various evolutionary scenarios. An environment may be more or less heterogeneous. A homogeneous environment exerts a consistent message concerning fitness, whereas a heterogeneous environment offers an ambiguous message. The structure of populations, too, can affect the consistency of selection: a more mosaic structure allows for more ambiguity. Gene flow is a homogenizing influence. As a rough rule, one migrant per generation will suffice to ensure genetic homogeneity. All of these factors matter tremendously for rates of divergence among populations. Any of these can affect the significance of selection, positively or negatively. So, with gene flow among populations, even selection for different traits will tend to be overcome by immigration. Smaller populations are, to take another example, more sensitive to the effects of chance, and gene flow will tend to reduce the impact of selection in a local population.

(5) *Trait polarity*   We need to know which traits are primitive and which are derived: we need to know whether a trait is ancestral within a clade, or whether its presence is the result of convergence among lineages. In general, we need to have an independently established phylogeny in order to know what is derived. A trait that is primitive within a group—that is, one that is present among ancestors—likely will not be an adaptation within the lineage. It will just be inherited. A trait that is derived within a group—that is, one that is novel among the descendants—is a better candidate for being an adaptation.[7] Current use or advantage is not enough to show that a trait is an adaptation. To remind us of an example from Darwin (1859), skull sutures are certainly adaptive in humans since they facilitate passage through the birth canal. They have an important, and even essential, current use. Skull sutures are nonetheless not an adaptation for parturition, because birds and reptiles also have skull sutures, even though they hatch rather than being born live. As Darwin says, even though these skull sutures "no doubt facilitate, and may even be indispensable for" mammalian birth, they are not adaptations for that function. *Adaptive features are not necessarily adaptations.* What is required to show that a trait is an adaptation, and to provide an evolutionary explanation for a trait in terms of natural selection, is knowledge of historical antecedents and conditions. Evolutionary history is the substance of adaptation.

Once again, these are five ideal conditions for adaptive explanations. An ideal explanation would offer sufficient evidence in each of the five categories. Real explanations often fall short of this ideal standard. Sometimes we make do, reasonably, with less than the ideal. Still, we should not accept the claim that some trait is an adaptation for some function without substantial evidence in support of the claim. Perhaps some pieces might be missing, but these conditions at least capture the *sort* of evidence that is relevant. We need not be troubled if a case falls short on some of the conditions, but we should be troubled if the evidence does not build a credible case. Think of it as akin to building a criminal case. We are only infrequently offered anything iron-clad in every way. Still, we do expect some aspects of the case to be made decisively. Perhaps we have clear opportunity and motive, but no witnesses. Perhaps we have witnesses, but no clear motive. We needn't have everything locked up for a conviction, but mere suspicion is not enough either.

The example Brandon relies on for support and illustration of the five constraints is the evolution of heavy metal tolerance in plants, which, as it happens, is a nicely documented case for adaptation in the biological literature (summarized in Antonovics, Bradshaw, and Turner 1971; and in Jain and Bradshaw 1966). Let's see how the example works and how it illustrates the five conditions. Grasses that grow on soil contaminated with heavy metals, such as the tailings from mining sites, often evolve a tolerance for the high level of metals in the soil. Without that tolerance, they die within the tailings, or at least have dramatically reduced viability. These locations, it turns out, not only have an overabundance of heavy metals, but are also low in the nutrients plants need for healthy growth. A surprising number of grasses can adapt to the severe conditions these sites offer. Let's look at how each of Brandon's five conditions works out in the case of heavy metal tolerance.

(1) *Selection*   In the case of heavy metal tolerance, the intensity of selection is very high, sometimes apparently as high as 95 percent (Jain and Bradshaw 1966). This means that the mortality of intolerant forms is near ten times that of the tolerant forms on the contaminated sites. This is intense as selection values go. Since the selection is intense and continual, it can be observed relatively directly. It is often much more difficult to measure selection than it is in the case of the grasses. It is not enough in general to know just that there is selection. It is also important to know the strength of selection pressures for the populations under study and the kind of variation that is present. The strength of selection is expressed, as above, in terms of selection coefficients or selection differentials. When selection coefficients are low or variable, selection can be hard to detect in the field. When they are high, as in the case of heavy metal tolerance, they are evident.

(2) *Ecological factors*   Specific mechanisms are required to render heavy metals nontoxic. (Think of the effects of lead paint on humans, who are not tolerant.) The advantage of tolerance is clear enough. In the case of heavy metal tolerance, just as intolerant types perform poorly on contaminated soils, tolerant types perform poorly on better ground. The cause for this difference

in performance is not exactly clear. In the contaminated soils, lower nutrient levels are likely to be important, as is the need to sequester the heavy metals. There is more than one reason the tolerant forms might perform well. On better ground, the intolerant grasses are likely to be more efficient in utilizing the resources available in uncontaminated soil. This shows that the tolerant forms are not merely more efficient in using the resources available. Whether we know exactly how the physiological mechanisms work or not, we do know the ecological factors that are responsible: the concentration of heavy metals makes the difference.

It is often difficult to identify the exact ecological factor responsible even when there is selection. In many cases, it is possible to show that selection occurs, but very difficult to provide a convincing mechanism. When that is so, we can meet the first but not the second condition. This is particularly common when there are few cases to observe. Discovering the pattern is difficult because many factors could affect fitness in any given case, and with fewer cases we cannot distinguish which factors are responsible. Sometimes, for example, a predator leaves a visible residue, and sometimes there is an obviously identifiable cause. In other cases, the cause is less clear. It is, though, *sometimes* possible to identify the cause in a way that is unambiguous. Heavy metal tolerance is one such case. Resistant forms grow better on contaminated sites than do nonresistant forms; and, correlatively, nonresistant forms grow better on uncontaminated sites than do resistant forms. Given the physical environment, the selection is clear. Antibiotic-resistant bacteria are another obvious case: resistance followed the discovery and use of penicillin, and it is clear that the administration of penicillin was the factor selecting for resistance.

(3) *Heritability* In the case of heavy metal tolerance, the genetic basis for the tolerance is not well understood, but it is clear that heritability values are relatively high (MacNair 1979). That is enough to guarantee that there will be a response to selection. With intense selection levels, the response can be significant even if the heritability levels are relatively low. With low selection levels, the response can be significant if heritability levels are high. Estimates of heritability are, again, difficult to obtain with any precision. Heritability is a statistical measure of the similarity between parents and offspring in a common environment (see Brandon 1990, ch. 2). If the environment is highly variable, and the variations in environment are correlated with differences in genotype, then the natural variation gives us no measure of heritability. This is a simple consequence of the well-known fact that heritability of a trait can change with the environment (Lewontin 1974a). In other cases, we can empirically measure changes in a population in some phenotypic or genotypic

trait—that is, the response to selection—but cannot tell what the intensity of selection is without some measure of heritability; and without knowing the intensity of selection, we cannot determine the heritability.[8] In this case, the heritability is high, and selection is high. The evolution should be rapid.

(4) *Population structure*   With wind-pollinated plants such as grasses, it is possible to get a relatively complete picture of the population's structure, since we can chart the pollination patterns. In one of the grasses that develops heavy metal tolerance, *Anthoxanthum*, different strains have different flowering times. This has the effect of isolating intolerant types from tolerant forms, which enhances evolutionary divergence. Many things are included under the broad rubric of population structure. For example, it is important to know the relevant scale so that sampling does not occur on a larger scale than an effective population (that is, the breeding population). It is important to know immigration and emigration values, so that the rate of change due to selection is not confounded with other sources of change. Low values for immigration facilitate evolutionary divergence; higher values inhibit it. Given the isolation due to flowering times, this is reasonably well understood. Without isolation, crossbreeding would slow or eliminate evolutionary change.

(5) *Trait polarity*   In the case of heavy metal resistance, because the sites are of relatively recent origin, it is unproblematic to identify the ancestral state; indeed, it is possible to identify a population that is not much different from the ancestral population. The ancestral populations were generally not heavy metal resistant. To use a slightly different example, one interesting feature of caste structure in social insects has received a convincing resolution in recent years in terms of kin selection. The key to the explanation of sterile castes depends on a haplodiploid mechanism of sex determination. It is clear that sterile castes have evolved several times within the *Hymenoptera* (ants and bees) from ancestors that did not exhibit sterile castes; this fact not only shows that caste structure is derived, but raises the likelihood that it is selected for.

As I've said repeatedly, these five conditions provide no simple litmus test for adaptive explanations. They can be thought of as illustrating how complete an explanation in terms of natural selection might look, were it ideally supported. The ideal explanation would have decisive evidence in all five categories. The case of heavy metal tolerance in grasses is nearly ideal. Again, most actual explanations will fall short of the ideal. Some explanations will fall short on many of the conditions, and often the evidence we have is indirect, but they may still count as reasonably good evolutionary explanations. Insofar as it falls short on one condition or another, an explanation will be problematic. There is no easy line to be drawn, marking off good from bad explanations, and there is room for difference of opinion. The explanation of heavy metal tolerance at least shows that it is possible to come close to satisfying all the conditions in a field study. This is enough to show that Brandon's criteria are not overly demanding. That is, it is not unreasonable to expect these conditions to be satisfied for adaptive explanations. This is, after all, an evolutionary explanation that does meet them. In the next section of this chapter, I'll illustrate the five conditions in more detail with another case, also a natural case that illustrates the conditions almost perfectly. It should reinforce the conclusion that the standards are not unreasonable.

Thus Brandon's five conditions capture, more or less, what would count as a good adaptive explanation in terms that are natural to population genetics. In the previous chapter, I discussed another approach to understanding the construction of adaptive explanations, based on reverse engineering. In the next chapter, I'll turn to a third method, based on descent. Although there are other methods, population genetics offers one very important approach to constructing adaptive explanations. Among biologists, it is a common method. Believing in adaptive explanations does not make one an *adaptationist*, of course. One might believe that there are some cases of adaptive explanations, but be skeptical about whether adaptive explanations are very common. Brandon does not suggest that his conditions are broadly met, though he is insistent that they are sometimes met. Though adaptationism is often discussed in terms of engineering design, rather than dynamic models, it's useful to consider what might facilitate or impede adaptation, and what adaptationism might come to, in these terms. Again, the root problem is one of methodology.

This will take us away from a focus on the evaluation of adaptive explanations and back toward the commitment to adaptation*ism* as a program of research. Since much of the discussion is carried on at this level among both defenders and critics of evolutionary psychology, we should at least see how to focus it in the current context. *Adaptationism*, Steven Orzack and Elliot Sober (2001, 6) tell us, is "the claim that natural selection is the only important cause of the evolution of most nonmolecular traits and that these traits are locally optimal." Optimality assumes that there are no constraints that inhibit selection from achieving optimal phenotypes: organisms are as good as they could be. Orzack and Sober contend that optimality guarantees the sufficiency of natural selection as an explanation of evolutionary dynamics, and that adaptationism requires optimality. These are issues Orzack and Sober have discussed in many places (1994a, 1994b, 1996, 2001). As they develop it in "Optimality Models and the Test of Adaptationism" (1994a), if we consider a specific trait, then there are three distinct propositions to consider. In order of increasing strength, these are:

(U)   Natural selection plays some role in the evolution of the trait.

(I)   Natural selection is an "important cause" in the evolution of the trait.

(O)   Natural selection is a "sufficient explanation" of the evolution of the trait, and that trait is optimal.

Heterozygote superiority—the case in which a genetically heterozygous individual has greater fitness than either of the genetically homozygous individuals—is sufficient to show that the last of these (O) is stronger than either (U) or (I). The heterozygous individual has two different alleles, whereas the homozygote has two genes of the same type. Conventionally, Aa is the heterozygote, while AA and aa are homozygotes, since genes are paired. If the heterozygote has the highest fitness, then natural selection certainly may be important in understanding the population dynamics; however, natural selection then will not be *sufficient* to explain the distribution of genotypes or phenotypes, and the optimal trait cannot be fixed in the population; homozygotes will have a reduced fitness and will occur in the population. So the most fit phenotype, the heterozygote, cannot be the only phenotype.

The evolution of malarial resistance in West African human populations, again, is a textbook case in which heterozygote superiority is appealed to in supporting and illustrating natural selection in human populations. As it is typically described, heterozygotes for the S (sickle cell) and A (normal) alleles gain an increased resistance. In this case, S is the recessive allele, and A is dominant. The parasite suffers from reduced fertility in heterozygotes. Homozygotes for the sickle cell allele S have a severe hemolytic anemia, usually resulting in early death. The problem for these SS homozygotes is a hemoglobin protein that disrupts red blood cell structure, especially in the absence of oxygen. Homozygotes for the dominant A allele are more susceptible to malaria. So AA homozygotes are at a disadvantage, as are SS homozygotes. The result is balancing selection with $f(A) \approx 0.89$ and $f(S) \approx 0.11$ for both predicted and observed values.[9] Natural selection may be an important factor without being the *only* important factor in the evolution of a trait, as it is in this case. Adaptationism correspondingly commits us to more than just the idea that natural selection plays *some* role in the evolutionary process. This last claim, (U), is uncontroversial.

Adaptationism likewise involves more than just the idea that natural selection is *important* in the evolutionary process. (I) is only marginally less controversial. The reason is that in cases like the sickle cell case, we do not have a purely adaptive explanation. There are other factors at work, even though natural selection is an important factor. Adaptationism, Orzack and Sober insist, commits us to the view that natural selection is generally sufficient to explain the evolution of a trait. So more cases should look like heavy metal resistance than like malarial resistance. Optimality (O) commits us to the view that natural selection is the *only* important cause and that the traits produced are locally optimal. Adaptationism in Orzack and Sober's hands then becomes the claim that (O) is at least generally true. Of course, that does not commit them to being adaptationists. Orzack and Sober are less interested in defending adaptationism than in explicating what adaptationism might mean. They are making a claim about what adaptationism consists in, what it would commit us to holding.

In a similar vein, Peter Godfrey-Smith (2001, 336) offers the following gloss of one of three kinds of adaptationism he considers:

Empirical Adaptationism: Natural selection is a powerful and ubiquitous force, and there are few constraints, except general and obvious ones, on the biological variation that fuels it. To a large degree, it is possible to predict and explain the outcome of evolutionary processes by attending only to the role played by selection.

This is very nearly, but not quite, equivalent to Orzack and Sober's version of adaptationism. It at least is committed to the view that natural selection is dynamically sufficient in a wide range of cases. Godfrey-Smith usefully distinguishes this claim from two others with which it is often confused. What he calls "explanatory adaptationism" emphasizes that natural selection is crucial to explaining the apparent design, or adaptedness, of organisms to their environment, and underscores the centrality of this task to evolutionary biology. This claim is, so far, silent on how commonly organisms are adapted to their environments. "Methodological adaptationism" insists that from a methodological perspective, it is useful to begin with the assumption that a trait is an adaptation; we can move to nonadaptive explanations, he says, only when all adaptive explanations are disproven (Mayr 1983). These latter forms of adaptationism are, as Godfrey-Smith says, importantly different from the empirical version. Most obviously, they could be true even if the empirical claim is false. Sometimes, advocates are clear in embracing some sort of methodological adaptationism, while at least remaining reticent about the empirical form. John Alcock (2001, chap. 4), while hardly a timid adaptationist, promotes the methodological virtues of adaptationism, emphasizing their testability; yet, he clearly acknowledges that this hardly requires some form of panadaptationism. But in the disputes over the viability of adaptationism, as Godfrey-Smith illustrates, advocates such as Dennett and Dawkins move freely between these varied claims. As Godfrey-Smith points out, this does not help the discussion.

Orzack and Sober differ from Godfrey-Smith specifically over whether adaptationism requires optimality. Orzack and Sober think it is important that adaptationism is tied to optimality: natural selection does, after all, tend to maximize fitness, and so if natural selection is the only important cause for their evolution, those traits will tend to be optimal. Somewhat surprisingly, Orzack and Sober eschew this simple line. Nonetheless, as they point out, "the biology *practiced* by most adaptationists is in fact, rightly or wrongly, a biology in which traits are explained as optima" (Orzack and Sober, 2001, 6). The key difference between Godfrey-Smith's empirical adaptationism and the adaptationism of Orzack and Sober can be brought out by distinguishing two components to (O) in Orzack and Sober's analysis:

(S)   Natural selection is a "sufficient explanation" of the evolution of a trait.

(H)   The population is homogeneous, allowing "no significant differences" among individuals in the fit of their phenotype(s) to the predictions(s) for the trait.

The optimality condition requires trait homogeneity (Orzack and Sober, 2001, 8; see also Orzack and Sober 1994a, 368–369): there should be little variation among individuals, and each should fit the optimal prediction. Godfrey-Smith is clear that his empirical adaptationism imposes no such requirement.

Orzack and Sober ask us to consider a case in which natural selection does cause the average value for some trait within a population to match the predictions of an optimality model. Suppose, for example, that within a population optimality requires some definite sex ratio, and that we find that prediction is met on average. Yet suppose there is variation among individuals, so that, although the mean sex ratio is optimal, individuals do not produce the optimal ratio. Perhaps no individual actually produces the optimal sex ratio. Orzack and Sober (2001, 8) say that "A claim that each of these traits is optimal in a different way would not be appropriate because a causally meaningful claim about optimality is a claim that natural selection governs the fate of competing traits." Even in a case of frequency-independent selection, it would be astonishing to find that a natural population does *not* differ from what would be predicted by a model that assumes, among other things, the absence of mutation, migration, and drift. (If we did not find a difference, that would suggest that the test simply lacked discriminatory power.) On Orzack and Sober's view, that would tend to disconfirm the adaptationist scenario. Peter Abrams (2001), in an essay that should be required reading for anyone engaged in the issues raised by optimality and adaptationism, points out that the prevalence of frequency-dependence raises a fundamental problem for the commitment to trait homogeneity. Frequency-dependence entails that the dependence of fitness for a phenotype depends on the relative abundance of a trait within a population. It is well known that this can preclude maximization of even mean fitness, as simple prisoner's-dilemma sorts of cases guarantee. John Maynard Smith's (1982) emphasis on game-theoretic models is a powerful way of representing the importance of frequency-dependence. If we assume there is a negative frequency-dependence, then the fitness of some variant will decrease as its frequency increases in a population. The population accordingly may reach an evolutionarily stable optimum without trait homogeneity. Orzack and Sober are, naturally, aware of such possibilities but deny these are evolutionary optima. However we decide to use the term, and whether or not we want to count these as evolutionary optima, these possibilities are an important category of cases in which the population is at an optimal average value, while there is individual variation.

Many opponents have argued that optimality will fail when factors other than natural selection affect the evolutionary dynamics (see, e.g. Gould and Lewontin 1979; Kitcher 1987; Lewontin 1987; Parker and Maynard Smith 1990). This is certainly true. Since no population is unfettered by mutation or genetic drift, and none is free from historical contingency, it is unlikely that populations generally would be exactly at the optimum. It is even more unlikely that there would be trait homogeneity for these values, given the residual genetic variation in natural populations. Moreover, as Lewontin emphasizes, organisms are complex interconnected units, gene interaction is common, and evolutionary trajectories often depend on historical contingencies. Abrams is very clear in explaining why we should not expect optimality models even to capture the reduced standard. Here is Lewontin's own summary:

> Exact optimality seems unlikely and may never occur. Even if it did occur, we would not be able to recognize it because of lack of knowledge of the full set of selective consequences of a given trait. Near-optimality (adaptation) probably occurs frequently, but it does not occur all the time, and we don't know the exact frequency with which it occurs. (Lewontin 1987, 285)

With positive frequency-dependence, a variant will increase or decrease in fitness as the frequency of the variant increases. Territorial displays among animals, for example, are effective only if they are common. The expected result would be multiple adaptive peaks, each relatively impervious to invasion. Which peak a population reaches will likely depend on variation in initial conditions. No single state will be optimal.

Let's set this issue aside for a moment and pursue the methodological question: How would we tell if an adaptive explanation is the right one? As I've said, some, such as Ernst Mayr, think we should assume they are correct until they are disproven. Adaptation becomes virtually a methodological *fait accompli.* Mayr was once less insistent. In his classic work, *Animal Species and Evolution,* he wrote:

Each local population is the product of a continuing selection process. . . . It does not follow from this conclusion, however, that every detail of the phenotype is maximally adaptive. If a given subspecies of a ladybird beetle has more spots on the clytra than another species, it does not necessarily mean that the extra spots are essential for survival in the range of that subspecies. (Mayr 1963, 311)

Mayr was a brilliant naturalist, for whom such observations seemed little more than observations. But even in *Animal Species and Evolution,* such observations some times yield to the theoretical commitment to adaptation.

Alcock seems comfortable with a more adaptationist view. This evidently assumes too much since we can always construct an adaptive explanation to replace one that has been disproven. That is one of Gould and Lewontin's complaints, that alternative adaptive explanations can always be constructed when one fails empirically, making adaptationism as a broader assumption true by default. It is evidently true that although specific adaptive explanations can be empirically tested, the evaluation of adaptationism in any of its forms is another matter. Gould and Lewontin acknowledge as much, complaining not that specific hypotheses are not tested, but that the assumption that traits are adaptations is not itself tested. In one case they cite, David Barash (1976) had suggested, somewhat tentatively, an explanation for the aggressive behavior of mountain bluebirds. Gould and Lewontin complain that Barash relied only on "consistency" with adaptation as sufficient to support his particular adaptive explanation and that nothing was done to test the assumption that the trait was adaptive at all. Alcock devotes considerable energy to criticizing Gould on the point, insisting that although the test used by Barash was inconclusive, it was certainly a test, and the hypothesis was testable. Alcock is right about the specific point, but misses the more fundamental issue Gould and Lewontin raise, namely, that adaptationism itself is an empirical claim and that testing specific adaptationist hypotheses is not the same as testing adaptationism as a general assumption.

Orzack and Sober think of the problem adaptationism raises in terms of causal models. The key condition constitutive of adaptationist explanations is that natural selection must be the only "important cause" or must be a "sufficient explanation" for whatever trait we are interested in explaining. This does *not* mean simply that natural selection is more powerful than other factors, such as drift or pleiotropy, in evolution. I am actually doubtful that it's possible to make much sense of these latter claims. In any case, that is not part of the project. Orzack and Sober's adaptationism is instead a claim specifically about the power and ubiquity of certain explanatory models. As Sober (1987,

116) says in another context, "Adaptationism concerns the power of certain simple models *of* selection; it is not a claim about the power of selection *in* evolution." Orzack and Sober (1994a, 364) are even more explicit in saying adaptationists emphasize the importance of natural selection: "Adaptationists do not deny that factors other than natural selection played some role in evolution. However, they believe that these other influences may safely be ignored."

Here is how they think of the case. In constructing adaptive explanations, we begin with "censored" adaptive models. These are models in which the only force represented is natural selection. The models include no other significant evolutionary factors; for example, mutation pressure, genetic drift, and gene interaction do not significantly feature in adaptationist models. The only question we then need ask is whether these censored models fit the data well enough to be supported. If some censored model explains the phenomena, then Orzack and Sober would conclude that that confirms the role of natural selection in that case: "Natural selection here provides a sufficient explanation because taking other factors into account could not significantly enhance the predictive accuracy of the optimality model" (Orzack and Sober 1994a, 363). Adaptationism as a program is correspondingly vindicated if natural selection is a sufficient explanation for most traits in this sense and if those traits are locally optimal.

This invites the objection raised, for example, by Brandon and Rausher (1996) that Orzack and Sober's approach is too one-dimensional. As described so far, if *some* censored optimality model is sufficient, then adaptation is vindicated as the cause for whatever trait we are interested in explaining. Brandon and Rausher object that so far as this goes, there could be an equally adequate censored model that would also be sufficient by this test, though that model might include no role for natural selection. We would then have two models, each empirically supported and capable of explaining the phenomena we observe. Orzack and Sober are not unaware of the possibility and say simply that "additional data or analyses are needed" in such cases (Orzack and Sober 1994a, 364). This raises a number of issues I cannot explore here. Godfrey-Smith usefully suggests that we might break the impasse by considering alternative models of roughly comparable complexity. Empirical adaptationism is then confirmed, according to Godfrey-Smith (2001, 344–345), "if, in the majority of cases a better fit to the data is achieved by a selection-based model than is achieved by any other model of comparable complexity." I want to suggest the modest point that this illustrates the comparative character of evolutionary claims. We need competing claims with different assumptions. Only then do we get evolutionary tests.

So here is the adaptationist claim, according to Orzack and Sober: Natural selection is the only "important" cause in the evolution of some trait provided that mutation pressure, drift, or genetic and phylogenetic constraints did not affect the evolution of the trait in the past. Orzack and Sober insist that evaluating whether a trait is optimal is a question of whether there is a quantitative fit between a model and observed traits: an optimal trait is the best among the competitors, where what is "best" is what exhibits the highest fitness. In the previous chapter, we saw some cases in which a quantitative fit was evident enough; this is especially true in the case of the water strider described there. Commonly, we find a qualitative fit at best. This is true, for example, in predictions of optimal foraging models. Again, we saw an example of that in the work of MacArthur, described in the previous chapter. Herre, Machado, and West describe work on the evolution of sex ratios among fig wasps, noting that sex allocation certainly has a direct and significant effect on fitness. They show that sex allocation broadly fits the expectations. There is a bias toward females, as we would expect under conditions of haplodiploidy; moreover, the bias decreases, as expected, with single foundresses, and inbreeding increases the female bias (Herre, Machado, and West 1991, 192–199). Nonetheless, the fit is qualitative at best. The sex ratios with more than one foundress, in particular, show more female bias than would be expected (ibid., 201). Orzack and Sober's requirement of a quantitative fit is a stringent one, not satisfied in this work. At least this shows that it would be wrong to complain that Orzack and Sober's adaptationism is too easily satisfied. It may in fact be very difficult to satisfy it.

This is useful as far as it goes, but it does not go far enough. The key inductive move is the semantic ascent: Orzack and Sober concern themselves *only* with the adequacy of models, with whether a *model* is adequate and not whether an adequate model is actually explanatory. The fact is that an adequate model—one that predicts the phenomena—may not offer the right explanation. They claim that if a censored model is able to explain the phenomena, then that is good, and perhaps sufficient, reason to conclude it captures the causal factors involved. The core problem lies with the implications of censored models. The shift to censored models allows Orzack and Sober to say that *for explanatory purposes* everything but selection can be safely ignored if adaptationism is right. This certainly fits what adaptationists say. For example, if someone objects that pleiotropy is a limit on the power of natural selection—if, that is, some gene affects more than one character—then the adaptationist response is that pleiotropic effects can be corrected by selection. As Richards Dawkins (1982, 35) puts it, "[if] a mutation has one beneficial effect and one harmful one, there is no reason why selection should not favour modifier genes that detach the two phenotypic effects." The assumption is not only that all mutations are possible, but that they would be expected; and that would provide the material for natural selection to do its work. If we assume that, then we have a guarantee that no possibility is inaccessible, as some manifestly are over the short term with a finite reservoir of variation. Moreover, if selection is weak, any linkage disequilibrium will be unimportant over the long run. Under these conditions, adaptationism might be right. The conclusion I would draw is that except under the most limited and idealized conditions, selection is *not* sufficient to explain observed changes in genotypic or phenotypic frequencies. The key question is how realistic these conditions are, or whether they are present in a given case. The mere possibility of constructing a model that relies on adaptation and ignores the other factors that might affect evolution is not enough to show that that model is one we should endorse. This is, really, nothing more than a transposition and generalization of the moral from the previous chapter. Models are free; Explanatory models need to have empirical warrant.

The issue raised by Brandon and Rausher, which I've echoed, concerns the converse of what I've called *semantic ascent*. They point out that it is not true that if there are censored models in which everything but selection can be safely ignored, then adaptation is the right explanation. It is not true that if there is an adequate censored selectionist model, then selection is the actual explanation for whatever effect we observe. Suppose Orzack and Sober are right, as they evidently are, in allowing that natural selection almost always matters for evolution, as do variation, drift, linkage, and other factors. Genetic and population structure are always causal factors, even assuming natural selection is capable of indefinitely modifying phenotypic structure. Consider cases in which censored selectionist models are *not* sufficient. If genetic and population factors are essential in explaining some phenomenon, then a censored selectionist model will not be enough. So, for example, in cases involving heterozygote superiority a censored selectionist model is *not* sufficient. The evolutionary trajectory and the equilibrium state in this case depend on the diploid genetics as well as the history of selection. Similarly, if there are significant nonadditive factors, then the evolutionary trajectory and the equilibrium will depend on these interactions. Selection alone will not be enough. If this is so, then there are comparable factors that are equally important in the cases in which censored selectionist models are empirically adequate. The models are censored, but they depend on assuming, for example, additivity of genetic effects; additivity is as important to the model as is linkage or dominance in others. Likewise, if a population is panmictic (that is, if mating is random), that will enhance the role of selection. It would be wrong to assume that only selection matters in such cases, or that population structure, say, is not a relevant causal factor. Random mating may be common or rare, but it is not universal. Even if the censored adaptive model does adequately explain the phenomena, that will hardly make natural selection a "sufficient" explanation. The causes are varied and complex even if the models are simple.

The morals here are parallel to those we drew in the previous chapter concerning reverse engineering. The mere possibility of constructing an adaptive explanation is not sufficient to show that that is the right explanation. It is of course not irrelevant since it does at least show the explanation is consistent with adaptation. Given the number of parameters available in these models, as expressed in our five constraints on adaptive explanations, there is a great deal of flexibility. Selective values can vary, as can assumptions about the extent and kind of variation. There may be many environmental parameters to consider, and population structure is highly variable. To make the point explicit, if we *assume* that some trait has high heritability, or that the population is panmictic, that would enhance the prospects for an adaptive explanation. But it is another question whether heritability is high, or what the breeding patterns might be. As we saw in discussing reverse engineering, the core question is not whether we can construct *some* model consistent with selection, but whether a proposed model is empirically motivated, whether there is evidence to support the claim that the proposed model is right. Constructing a censored adaptive model that is empirically motivated, as Orzack and Sober require, is not an easy task. Our knowledge of the relevant parameters may be scant or incomplete, in which case we might not be in a position to know whether some proposed model has sufficient empirical warrant to merit our assent. Even in that case, the possibility of constructing a censored adaptive model does not show that there isn't an equally defensible nonadaptive model. In the end, the explanatory problem is not merely one of fitting some adaptive model to the data, but comparing alternative explanations. The first question that concerns us, at this stage, is how the adequacy of these adaptive models is addressed. When we have that in focus, we can then ask about the adequacy of the explanations offered within evolutionary psychology. The question of how these models should be evaluated can best be seen in an example, once again structured around the five criteria for adaptive explanations.

3   Regressive Evolution and Natural Selection

There are many convincing dynamic studies of "natural selection in the wild" (Endler 1986). One I find especially compelling concerns the role of adaptation and natural selection in cave environments. A remarkably similar suite of characteristics can be observed in a wide variety of contemporary forms that are limited to caves. Most species exhibit some elaboration of certain structures relevant to perception, such as antennae in the case of many invertebrates and the lateral line system in the case of some fishes, which facilitate motion detection. There is characteristically an elongation of the body, including appendages. But the hallmark of cave-dwelling life forms is doubtless the reduced or lost features, in particular, the loss of eyes and pigment. This suite of morphological features has been termed *troglomorphy*. Historically, very different evolutionary explanations have been offered for the elaborated as opposed to the reduced traits. This is fairly natural if we assume that complex features should be explained in terms of natural selection. One set involves enhanced features, the other reduced, or regressed, features. Elaborated features appear "more obviously" adaptive, in that they presumably increase the organism's ability to survive and reproduce in the cave habitat. They often involve nonvisual sensory structures such as antennae, taste buds, or tactile senses; they arguably provide sensory compensation in an environment in which the visual senses would be wholly useless. The reduction or loss of features, often referred to as regressive evolution, is less obviously adaptive.

The discussion of the origin of regressive features in cave animals dates to some of the earliest theorizing on evolution. In one of the first systematic defenses of evolution in biology, Jean Baptiste de Lamarck (1744–1829) took note of a strange salamander living in caves. It was colorless and blind.

The *Proteus*, an aquatic reptile allied to the salamanders, and living in deep dark caves under the water, has . . . only vestiges of the organ of sight. . . . Light does not penetrate everywhere; consequently animals which habitually live in places where it does not penetrate have no opportunity of exercising their organ of sight. . . . Now animals belonging to a plan of organization of which eyes were a necessary part, must have originally had them. Since, however, there are found among them some which have lost the use of this organ, . . . it becomes clear that the shrinkage and even disappearance of the organ in question are the results of permanent disuse of that organ. (Lamarck 1809, 116)

Lamarck thus explained the reduced eyes of *Proteus* as a consequence of their lack of use, as he did the reduced eyes of moles. He surmised that in a form that has fully developed eyes, complete lack of use will cause their eyes to be underdeveloped, or to degenerate. If these losses due to disuse and degeneration were then passed on to offspring, those offspring will likewise have reduced vision. Over time, Lamarck thought a repeated cycle of disuse, degeneration, and inheritance would lead to species such as *Proteus*, with only vestigial organs where their ancestors had eyes. This is *regressive* evolution in his

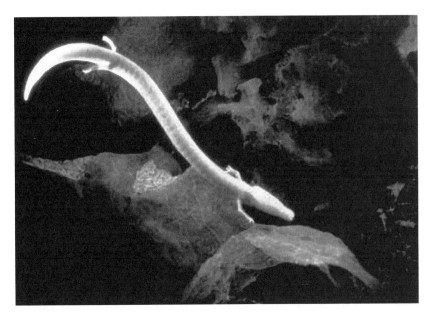

Figure 3.1
Proteus anguinus. This is a functionally blind and depigmented salamander, mentioned by Lamarck and Darwin. Both thought the regressed features were the consequence of disuse. *Proteus* is limited to portions of Italy and Slovenia. *Proteus* is extremely large for cave organisms. From http://vennarecci.free.fr/Rhinogrades/images/Page9/prot%E9us.jpg/.

view precisely because it involves the loss rather than acquisition of complex characters.

Darwin similarly took note of cave dwellers a half century later in his *Origin of Species* (1859). By this time, a number of additional cave species were known. Nearly all, like *Proteus*, are blind and lack pigment. One of the most famous was described by J. E. deKay in 1842 and later made famous by Louis Agassiz, certainly the first great American biologist. This is the blind fish of Mammoth Cave, *Amblyopsis spelaea*. Like Lamarck, Darwin attributed the loss of eyes in the fish to disuse. He says, "It is well known that several animals, which inhabit the caves of Carniola and Kentucky, are blind. . . . As it is difficult to imagine that eyes, though useless, could be in any way injurious to animals living in darkness, their loss may be attributed to disuse" (Darwin 1859, 135). As was generally the case, Darwin appealed to the effects of disuse only when no obvious adaptive advantage presented itself. Nonetheless, Darwin readily resorted to natural selection for the elaborated traits of cave animals. So, for example, enhanced sensory features offered evident advantages; here Darwin appealed not to the effects of use, but to natural selection. Darwin was, however, reluctant to ascribe any conspicuous role to natural selection in caves. He explains his reluctance this way:

Far from feeling surprise that some of the cave-animals should be very anomalous, as Agassiz has remarked in regard to the blind fish, the *Amblyopsis*, and as is the case with blind *Proteus* with reference to the reptiles of Europe, I am only surprised that more wrecks of ancient life have not been preserved, owing to the less severe competition to which the scanty inhabitants of these dark abodes will have been exposed. (Ibid., 136–137)

One reason Darwin favored disuse as explaining the loss of vision is due to the thought that these are, as he says, just "wrecks of ancient life," and thus not very revealing about the adaptation of organisms to their environment. Another is that Darwin thought that competition fueled natural selection and that the cave environment was one in which competition was reduced. Natural selection could not do its work if there is no competition, and it cannot do it efficiently if competition is reduced.

The view that acquired characteristics can be inherited is of course no longer tenable, and so the "explanation" of regression or loss in terms of disuse is likewise untenable; nonetheless, the evolutionary explanation for regressive features is still controversial. One approach invokes neutral evolution, that is, evolution of features favored neither directly nor indirectly by natural selection (see Kane and Richardson 2005 and Wilkens 2005). On this view, the key assumption is that features such as eyes and pigmentation offer no advantage, and so are of no selective value; that is to say, neither their maintenance nor loss would affect the fitness of a cave-dwelling organism. Therefore, mutations that affect these features will not be removed by natural selection. Since complex organs such as eyes require many genes to develop properly, mutations on average will tend to be degenerative, and therefore the net effect over time will be the reduction or loss of these characteristics. This type of explanation has been offered to explain eye and pigment reduction in the Mexican cavefish, *Astyanax faciatus*, as well as in other cases. Alternative explanations for regressed features have invoked natural selection, usually indirectly, in the form of evolutionary trade-offs. For example, since caves are supposedly food-poor environments, many biologists have suggested that natural selection would favor energy economy. Thus, individuals that put less energy into the development and maintenance of "useless" structures, such as eyes and pigmentation, may survive better and produce more offspring than those that maintain such features. This suggests that regressive evolution can have an effect on fitness.

Hypotheses are of course one thing, and confirmation is another. More recently, one of the best-documented studies of natural selection in caves focuses on a freshwater amphipod crustacean, *Gammarus minus*. Much of this work was conducted by David C. Culver, Thomas C. Kane, and Daniel W. Fong (see Culver 1982; and esp. Culver, Kane, and Fong 1995). The work concentrates on a series of caves in West Virginia, each with a population of *Gammarus*. This work suggests that, at least in the case of *Gammarus minus*, the reduction and elaboration of features may act in concert to produce adaptation to the cave environment. Both reduced and elaborated characters are treated similarly and in concert. Let's see how this case fares in light of the five conditions we have been focusing on as constitutive of adaptive explanations (for a brief discussion, see Culver, Kane, and Fong 1995, chap. 9; also Kane and Richardson 2005).

(1) *Selection*   *Gammarus minus* oftentimes has dramatically reduced visual structures. They have compound eyes, with numerous eye facets, as do all insects. So reduction in visual structures can be simply counted, by comparing the number of facets, or ommatidia. Whereas surface forms of *Gammarus* average around forty ommatidia per eye, cave forms can average less than two. There are also reductions in the size of the optic lobe and an increase in the size of olifactory lobes. One of the advantages that *Gammarus* offers as a model organism is that it has invaded caves multiple times, and correspondingly its suite of troglomorphic characters has evolved more than once.[10] Another advantage is that there are related surface forms, which presumably are similar to the ancestors of cave forms. When Jones, Culver, and Kane (1992) measured selection, they found a consistent pattern of selection for larger body size, larger antennae, and smaller eyes was evident in the cave populations. (Antennal size and head size are likely not independent

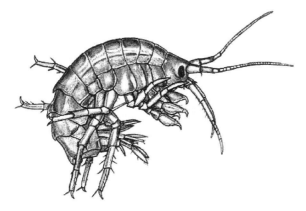

Figure 3.2
A depiction of a freshwater form of *Gammarus*. The toglomorphic version has elongated antennae, and much smaller eyes. From http://www.feketitsch.de/foxhouse/gammarus_roeseli.jpg/.

characters.) In the case of eyes, there was selection for smaller eyes within cave populations and selection for larger eyes in the surface forms.

(2) *Ecological factors*   Caves supporting biological communities can generally be described as food-poor relative to surface habitats and with constant temperature, humidity, and darkness in zones away from the entrance (see Barr 1968; Poulson and White 1969). These characteristics are both obstacles and advantages to cave-dwelling organisms and provide some of the selection pressures to which these species must adapt. Various evolutionary scenarios have been suggested for the changes characteristic of cave-limited species. One view suggests that colonization of caves may have been a passive process driven by climate changes in the surface habitats of the ancestors of present-day troglobites. Glacial advances and retreats have been posited as one such climatic change: species that were adapted to the cool, moist surface environments of glacial maxima were extirpated during warm, dry interglacials, leaving only remnant populations living in cave entrances and deep sinkholes that were still cool and moist. Whatever else, such a scenario would explain the isolation of cave forms and why there are often no related surface forms. These isolated populations subsequently underwent further adaptation to the cave environment, producing many of the current troglobitic taxa. Such a scenario has been suggested for the more than 250 species of cave-limited trechine carabid beetles in the United States, for example. Yet another scenario suggests that caves may have been actively colonized by ancestral stocks because of the advantages that caves afforded, especially the constancy of temperature and the relative lack of predators. The passive colonization view implicitly emphasizes the problems that incipient cave dwelling must solve, whereas the active colonization view explicitly emphasizes the advantages of cave habitats.

In the several *Gammarus* cave populations studied in West Virginia, the hydrological structure promotes the isolation of one population from another. As in other cave forms, the populations have reduced resources and live in total darkness. In many cases, though not in the West Virginia populations, *Gammarus* cave populations are not substantially different from spring populations. The evolution of troglomorphic features depends on the isolation of these populations from surface forms, and this generally requires relatively large caves. The actual selective factor behind selection for reduced eyes is uncertain, but there is a plausible candidate. One key component of fitness concerns the ability to reproduce. Field studies on several populations from caves have demonstrated that individuals with smaller eyes, larger antennae, and larger body size mate more frequently and produce more offspring than those in the same population with larger eyes and smaller antennae and body size. Therefore, with regard to eye size and antennal length, the more troglomorphic individuals in the population are more fit. The increased antennae size makes it easier to find mates, and reduced eye size becomes a trade-off for increased antennae size. Reduction is a trade-off. Chemosensory (taste) and optic (visual) inputs are integrated in the olfactory and optic lobes of the brain respectively. The brains of troglomorphic cave-dwelling individuals have greatly reduced optic lobes and greatly increased olfactory lobes relative to the optic and olfactory lobes of surface-dwelling individuals. The evolutionary need to improve chemosensory functions leads *Gammarus* to co-opt portions of the brain previously involved in the now useless visual function. This explanation is consistent with Darwin's view of sensory compensation but, unlike Darwin's explanation, supports the view that both antennal enlargement and eye reduction are the result of natural selection, producing adaptation to the cave environment.

(3) *Heritability*   Heritability is difficult to estimate in the field. It requires significant and sophisticated laboratory control. In the case of *Gammarus*, Daniel Fong (1989) raised captive animals in a laboratory breeding experiment. Fong found that the characteristic differences between cave and surface forms, when they were raised in settings with representative light/darkness and temperatures, were persistent. The offspring of cave forms had smaller eyes than the offspring of spring forms. This at least indicates that differences between surface and cave forms have a genetic basis. It turns out to be difficult to get good measures of heritability within populations (because breeding is difficult in the captive populations), but Culver, Kane, and Fong suggest that (broad-sense) heritability values are large, averaging 0.7 for eyes. This is roughly comparable to heritability values for human height, which are also high.

(4) *Population structure*   Characteristically, cave-adapted organisms occupy a very limited range—sometimes limited to a single cave—but with no closely related surface relatives. *Gammarus* is distributed in the central United States from southern Pennsylvania to eastern Oklahoma, usually in surface springs and spring runs. In the two karst areas that Culver, Kane, and Fong focus on— one in West Virginia and the other in southwestern Virginia—populations of *G. minus* also occur in caves. These cave populations are relatively large, sometimes comprising more than 10,000 individuals. Genetic data suggest that there is little or no interbreeding between cave- and spring-dwelling populations within these two areas. (Studies of protein variation were used to measure the amount of inbreeding.) The data clearly indicate that there is no gene flow

among the cave populations. They are distinct populations, which independently evolved reduced eye size.

(5) *Trait polarity* Trait polarity is readily settled in this case. The surface forms, as I've already said, have much larger eyes and do not lack pigmentation. Indeed, the cave forms follow the typical pattern for cave-limited forms, in terms of how they differ from the surface forms.

4 Natural Language and Natural Selection

This excursion into the evolution of some relatively obscure (but nonetheless interesting) organisms is meant, first of all, to enforce a single moral, namely, that the standards suggested for adaptive explanations are not set impossibly high. Secondarily, it illustrates how we can go about establishing what we need to know in order to validate adaptive explanations. Not only in the case of cave organisms, but in other cases, the standards for validating adaptive explanations that I advocate are actually met. These cases are impressive empirical studies of adaptation in nature. Let us see how well the proposals of evolutionary psychology fare against this kind of standard.

Steven Pinker and Paul Bloom have defended the idea that human language is the product of Darwinian natural selection. The principal argument offered by Pinker and Bloom is simple and elegant. It rests on the premise that adaptive complexity requires an explanation in terms of natural selection (Pinker and Bloom 1992, 454ff.). As Darwin said, without natural selection, we leave "the coadaptations of organic beings to each other and to their physical conditions of life, untouched and unexplained" (1859, 4). Human language certainly does have the marks of an adaptation; it is a highly complex capacity. More to the point, it involves a mosaic of capacities that are dependent on one another. We are able to extract phonemes from a rapidly changing auditory input and to coordinate movements in producing complex speech. Speech requires solving a difficult problem of motor control: producing speech depends on coordinating the face, tongue, and larynx, all while modulating breathing. We can access a wide and varied lexicon. We are able to encode and decode statements of staggering complexity with stunning efficiency. Human languages place enormous computational demands on the language user, and even more on the language learner. Language is evidently governed by rules and representations that are not taught in any straightforward way. There are rules governing phrase structure that provide information about the underlying structure and meaning of utterances; grammars depend on lexical and phrasal categories that simultaneously limit what can be said and give language its remarkable expressive power. Children acquire language with remarkable proficiency, without formal instruction—a fact that is most naturally explained by the presence of substantive prior constraints on languages that facilitate learnability (Chomsky 1972, 1980). Its universality among human societies alone is taken to suggest that it is part of our evolutionary heritage. This is precisely the kind of complex capacity that is plausibly the consequence of natural selection. Pinker and Bloom exploit these general features in concluding that language shows "signs of design" (1992, 459).

Language does seem to be a natural candidate for explanation in terms of evolution by natural selection. Do we have anything more substantive than this general suggestion? The problem is, after all, not merely to show that language is an adaptation, but to explain it as an adaptation. Without a suitable explanation, we will not know what specifically it is an adaptation *for*. In considering this question, we have 'an embarrassment of riches. As Terrence Deacon says, "looking for the adaptive benefits of language is like picking only one dessert in your favorite bakery" (1997, 377). Language use allows for communication, thereby facilitating coordination. This means it offers benefits for hunting, food sharing, toolmaking, social bonding, planning, mating, caring for offspring—anything that depends on social life. That is, for us, just about anything at all.

Let's look at how well scenarios offered for the evolution of human language fit the ideal type for adaptive explanations.

(1) *Selection* The very general case that Pinker and Bloom offer does support the claim that there has been selection for linguistic capacities. This tells us nothing about the strength of the selection or its character. It doesn't, more specifically, tell us which features of language were selected for directly and which only indirectly. It doesn't tell us which features are specifically tied to language and which have other sorts of evolutionary explanations. It is not even clear from the scant evidence we have whether language arose with modern humans or is older (see Lieberman 1991, 1992). If the origins of language are relatively ancient, perhaps following the enlarged brains of *Homo habilis* or *Homo erectus*, then maybe the complex suite of adaptations for speech would be intelligible, though they would no longer be derived traits within *Homo sapiens*. Perhaps only some of the features characteristic of *Homo sapiens* arose with the use of language. There is little disagreement that a trend toward enhanced brain sizes was present with *Homo habilis* and continued with *Homo erectus* (see figure 3.3). The functional significance of these changes is unclear, though it wouldn't be surprising, to say the least, if these early hominids had some sort of language.[11] Certainly, it would be shocking if they did not have complex communicative skills.

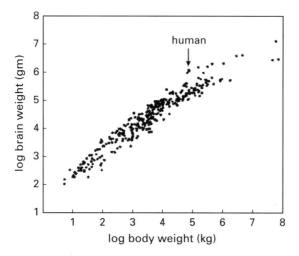

Figure 3.3
Changes in brain size as a function of body size. *Homo sapiens* has a larger brain than would be predicted from body size alone, ranging from roughly 1,200 to 1,700 cubic centimeters, or about 600 grams. We assume weight and volume are equivalent measures. *Australopithecus* is much smaller, and more nearly the primate norm. From Deacon 1997b.

The kind of information this gives us is woefully inadequate to the task set for it by evolutionary psychology.[12] Natural or sexual selection requires contrasting groups, differing in trait values. One group might have a trait, and another lack it. Alternatively, individuals might possess a trait to varying degrees. In either case, selection depends on prior variation and an explanation in terms of selection depends on knowing something of the character of that variation. Although there are some individuals differing in linguistic abilities in significant ways among contemporary humans, this provides no information concerning the natural variation in protolinguistic communities. An individual suffering brain damage is not one who somehow regresses to ancestral patterns. Furthermore, even when there are contrasting groups, selection differentials are often very small (a selective difference of 1 percent is an enormous evolutionary advantage). In the case at hand, we have no information whatsoever concerning the kind of variation or the effect it would have on fitness. We don't even know what hominid group would be appropriate to look at, though there is likely more than one. We thus have nothing substantive concerning the kind or character of the selection pressures influencing the emergence of language. That doesn't mean language wasn't subject to selection, or that we don't know a good deal about the evolutionary emergence of language; we do. The problem is that what we do know doesn't help much with the specifically linguistic capacities—syntactic, semantic, or phonetic—described by evolutionary psychologists.

The case is no better when we turn to comparisons with primates with the hope of understanding the selection or variation, largely because of the enormous behavioral, physiological, and cognitive differences between humans and other primates. There were early attempts to train primates to mimic human vocalization, though they met with little success. There were some attempts to explain these results away in terms of differences in the vocal apparatus, which are substantial, but it seems equally likely that the critical differences are in neural control mechanisms. The kind of complexity characteristic of human speech—again including syntactic, semantic, and phonological features—is, according to Pinker, a species-specific adaptation with underlying neural control or cognitive mechanisms.[13] Even here, the superficial comparative approaches are replete with difficulties. There are numerous anatomically distinct brain regions that are implicated in one way or another in language production. Typically, but not universally, these are in the left hemisphere. Damage to these regions results in a constellation of characteristic deficits, known and studied at least since the nineteenth century. What is called "Broca's area" lies anterior to the temporal lobes. Damage to it leaves individuals unable to make grammatical distinctions (for example, between active and passive voice), but still able to mark semantic differences. Wernicke's area lies posterior. Damage to it leaves individuals unable to mark semantic differences (for example, between different verbs), but still able to make grammatical distinctions. More recent studies using electrical stimulation and techniques such as fMRI and PET scans show that these areas as well as many others are involved in normal language use. Moreover, there are difficulties in characterizing with any precision the exact differences between the classical language areas. To make things worse, there are significant differences between individuals. Even the cause or significance of lateralization between the hemispheres is not known.[14]

(2) *Ecological factors*   What ecological basis do we have for selection *for* linguistic abilities? Doubtless, linguistic abilities are in one way or another an adaptation to social life. The relevant environment is the social community. The more specific answer we are offered is that language use facilitates communication. This is no doubt true; it is also nearly wholly uninformative. Communication can, after all, take many forms. Chimpanzees can certainly communicate with one another and with their human handlers. Their communicative skills were subject to natural selection. Dolphins utilize a distinctive set of whistles. Their use of these whistles was favored because it facilitated communication. Birds use a variety of visual displays in mating. These were

favored because they facilitated communication. Bees convey the location of pollen sources using ornate dances. These dances were favored because they facilitated communication. Ants communicate by laying down pheromone trails. The use of pheromones also was favored because it facilitated communication. No doubt each of these behaviors was favored because it "facilitated communication." This is just not enough. In order to explain why our linguistic abilities were favored by natural selection, we need a much more articulated hypothesis that answers a variety of questions. Some communication systems are relatively simple and stereotyped. Some are not. Human language is supposedly distinguished by its complexity. How do we explain those features? The appeal to communication, alone, will not explain the complexity of the human lexicon, say, or the dependence on phrase structure rules. The undoubted fact that language facilitates communication does not tell us about the features of human language that make it most remarkable, any more than the fact that the hand is used for grasping explains the complex structure of the fingers.

We should, again, not overestimate what we do know. Paleoecology has given us a good deal of information concerning the environment relevant to hominid evolution, reconstructed from such things as the distribution of extinct flora and fauna. We know, for example, that during the Pleistocene (4 million to 1 million years ago), australopithecines were widely distributed in Africa, occupying both woodland and savanna habitats; and there is reason to think that climatic changes were significant in the emergence of hominids. We know these australopithecines were bipedal; so we know our hominid brethren were likewise bipedal. The various species of *Homo* that subsequently developed occupied a variety of habitats, and, in spite of the considerable variation between them, their emergence is generally associated with the presence of tools. The first crude tools are coincident with the appearance of *Homo habilis*, some time between 2.0 and 2.5 million years ago. Such information offers scant help in understanding the factors that shaped the evolution of language, though they tell us a great deal about human evolution. Specifically, we need to answer questions like these: What factors in our ancestral environment favored verbal communication? Why did the complexities of human grammars arise? How were these factors different from those that shaped the communicative abilities of our prelinguistic and protolinguistic forebears? How did the linguistic abilities of *Homo sapiens* differ from the corresponding abilities in related groups in the genus *Homo*?

These are hard questions, not yet addressed in any significant way. It is in fact difficult to see what evidence would be relevant to resolving them. If we are eventually enlightened, that enlightenment will almost certainly not come from investigating human capacities, but from indirect evidence literally unearthed by paleoanthropologists. One salient fact must be obvious, though: If the relevant environment is the social environment, the sort of hard evidence favored by paleoanthropologists will not be likely to tell us what we need to know.

(3) *Heritability*   It might seem easy to answer the question of heritability. It is part and parcel of contemporary linguistic theory to emphasize the underlying unity of human languages and to downplay their diversity. Chomsky has spent a substantial part of his career arguing that "grammar" is innate (see, e.g., Chomsky 1972, 1980). He emphasized, simultaneously, the complexity in the structure of human grammars and the facility of children in acquiring languages. He once argued that there was an underlying "deep structure" to human languages, cutting across their superficial diversity. Deep structure has since disappeared from the system, but he still argues that the learning of languages could not be accomplished with generalized learning rules, given the conditions of human language acquisition. The underlying similarities that characterize human languages are supposed to be both universal and deep, providing explanations of both the similarities among existing languages and the ease of language learning. Given constraints that are both universal and contingent, Chomsky argued the only explanation would seem to be that these constraints are somehow innate.[15] All of this can be granted, but it does nothing to address the issue of heritability as that notion is deployed in evolutionary biology. Chomsky does not suggest it would. Heritability is a measure of the relative significance of genetic variance. It requires that there be significant variance to begin with, and if the variance is zero, then heritability is simply undefined.[16] What we would need is a measure of variance for linguistic competence and a measure of variances within and between different groups for ancestral populations, in order to assign heritability values. That is not addressed at all in the discussions of the innate constraints, whatever they might be, on language acquisition. It is something that is simply unknown. It is difficult to see how such values could be obtained, limited as we are to contemporary populations. Pinker (1994) similarly emphasizes the importance of innate constraints on language acquisition in *The Language Instinct*. Some of Pinker's most interesting psychological work has been concerned specifically with language acquisition. The universality and naturalness of our linguistic predispositions are exactly what lead Pinker to suggest there is a "language instinct." For all this matters, there might even be a "language instinct," but that is not the point. If the instinct is universal—if the constraints on language acquisition do not vary among individuals—then the heritability is undefined. Without knowing the variance in ancestral populations, we cannot even pose

the evolutionary question. The lack of (genetic) variance in contemporary populations does not help. And we don't know anything relevant about the ancestral groups.

Even this is probably too easy. Evolutionary psychologists often claim that heritability is unimportant precisely because the traits (supposedly) exhibit no variation. This would be simply shocking if it were true. Even for important psychological traits, human behavioral geneticists find significant heritability when they explore the question empirically. Likewise, evolutionary biologists find significant residues of genetic variation in natural populations. Additive genetic variance for such things as IQ, personality type, depression, and hyperactivity can exceed 40 percent (see McGuffin, Riley, and Plomin 2001). Among those who insist they are dealing with biological universals, we rarely see empirically grounded estimates. In any case, what we would actually need in order to build an empirical case would be estimates or values for specific variations in ancestral populations. Absent some information concerning ancestral variation, it's hard to know what to make of heritability claims. We might assume a more or less average heritability, but that might not be right for traits under persistent selection. Even if current variation were rich, that would tell us less than we would want to know about the sort of ancestral variation present.

(4) *Population structure*   What do we know concerning the structure of the environment and population relevant to gene flow? There is, of course, a great deal known about primate social organization. There is something known, too, about early hominids; for example, they were habitually bipedal, and the importance of tools and hunting to the earliest species within *Homo* indicates a level of cooperative behavior. Chimpanzees and gorillas also are cooperative and social. Early hominids were certainly social. Knowing our ancestors were social is one thing, of course; knowing anything about their forms of social organization is another. Knowing what aspects of this organization are relevant, and how, to language is yet a third. Beyond this we have speculation and extrapolation from contemporary groups and primates. We are left with very little firm knowledge of the social structure of early hominid societies. I will spend some time later discussing what we know of hominid phylogenies and what we know of the fossil record, but the simple point is that we know very little about the population structure of early hominids. We would naturally expect that hominid family groups would not be large. Clearly, selection did not favor traits because they allow us to live in large anonymous urban communities. But what would be the relevant structure? How large were our ancestral groups? Would there likely be outcrossing with other family groups?

Would there likely be matrilineal groups? Would males be likely to move away from their family groups? Would females? Would they be monogamous or polygamous? These are the sorts of questions that are critical to fixing questions of population structure. Among primates, the simplest rule is that everything is variable. The paleoanthropological and anthropological data suggest that variability is the rule. What would we predict? We do not know. Of course, that does not forestall confident conclusions, but it should.

(5) *Trait polarity*   It is at least easy to see that human language is a derived trait. Our primate ancestors had nothing approaching human language. The obvious problem is that we have little contemporary information that is helpful in reconstructing the ancestral traits since no traits that are specifically helpful to the evolution of language that are present in contemporary primate cousins shed any light on human abilities. We can do somewhat better by turning to the fossil forms, though the evidence is indirect. There are evident trends in brain size among our ancestors. Humans have larger brains than would be expected, even as primates. Primates have larger brains than would be expected, even as mammals. Australopithecines are nearer the primate average, and *Homo habilis* is, conveniently, an intermediate. Brain size is one thing, but it is yet another to address questions even about the size of cortical structures relevant to language. There are, naturally, increases in the size of neural structures typically associated with language as we move from chimpanzees to australopithecines to humans. The size of all cortical structures increases over these groups. Evidence concerning the cortical structures typically associated with language (e.g., Broca's and Wernicke's areas) is less than decisive. Human brains exhibit a trend toward lateralization, with language typically localized in the left hemisphere. (The right hemisphere is, though, involved in a variety of linguistic processing, such as processing rhythm and pitch.) Broca's area is a region involved in speech in most humans, but even among ourselves the superficial anatomical landmarks are variable. Even lateralization varies. It is clear that there were increases in the area among hominids; it is not clear what the functional significance of these changes would be.[17] The basic problem is located nicely by Terrence Deacon:

Why did language first evolve? In hindsight it is easy to justify as a useful means of communication, but the fact that it evolved only once gives us no comparative perspective. Hominid ancestors were the first to achieve the necessary mental conditions for language, but we do not know what these were. (1992b, 133)

Understanding how a complex trait evolves depends on knowing the primitive traits in ancestral groups that make its evolution possible. In this case, we are ignorant of what we would need to know in order to offer an explanation.

The overall assessment is clearly dismal. It is plausible to think that our linguistic abilities are the result of a history that involved selection for those abilities. However, in the face of the request for a substantive explanation of the character of the selective forces, their ecological basis, information about the heritability of the relevant traits, the genetic structure of the population, or the relevant ancestral traits, we are given nothing, save the facile suggestion that the evolutionary function of language is to facilitate communication. Again, this is almost certainly true, but is so general as to be wholly uninformative. There are many systems of communication, of which human language is only one. Moreover, the complex suite of characteristics that make it so plausible that human language is an adaptation is not explained by this very general characterization of the evolutionary function of language. How are we to explain the lexical and phrasal categories incorporated in human language? Dominance rules and precedence rules governing linear order are also part of human languages. These are the kind of features Pinker and Bloom appeal to in their "argument for design in language" (1992, 460ff.). Yet we are offered no explanation of these features. Pinker and Bloom reasonably emphasize the various traits that are necessary for human language, and they reasonably insist that these must be adaptations; but they offer no semblance of an adaptive or evolutionary explanation for the emergence of human language.

5   Human Reasoning and Natural Selection

There are two broadly different kinds of proposals that have been offered concerning the evolution of human reasoning. The first, exemplified by Robert Nozick, suggests that the evolutionary function of human reasoning is that it somehow enables us to deal with new and complex situations. We are creatures evolved to fit a variable environment, at least in terms of the physical demands and perhaps in terms of social demands. Rationality would then be appropriately treated as a more-or-less general-purpose capacity, applied across widely differing domains. Dennett's views are similar in tenor.

The second approach, exemplified by Cosmides and Tooby and embraced by both Gerd Gigerenzer and Matt Ridley, suggests that the evolutionary function of human reasoning lies in specialized adaptations; for example, its origins might be found in social exchange. Social exchange, including the sharing of food and other resources, was an important part of our evolutionary heritage, and Cosmides and Tooby claim that the result is an evolved set of mechanisms specialized for dealing with social exchange. Rationality, on their view, would be appropriately treated as a special-purpose capacity, or set of capacities, activated in different domains. These are provocative claims. I will bypass, for the moment, issues about the psychological and behavioral evidence for the alternative views. Let's see how they fare as explanations of human reasoning, considered as an adaptation.

(1) *Selection*   There is no doubt that there has been some selection favoring human rationality, or at least our cognitive abilities. Interspecific comparisons of brain size suffice to drive the point home. The most natural and primitive measure of brainpower would be gross size. That turns out to be less than satisfying since large ungulates and elephants have larger brains than humans, measured in terms of gross size. The mathematically sophisticated response is to scale brain size against body size. Allometric adjustments—which require comparing gross size to other growth factors—do lead to a more satisfying result: a plot of brain versus body size yields a relatively consistent slope of approximately 0.75 on a log-log scale. So as we see an enhanced body size, we should see an enhanced brain size, and the scaling tells us how much larger that would be. (See figure 3.3.) There is a remarkable amount of consistency in such measures and, in general, it is possible to predict brain size reliably from body size (see Deacon 1992a). Primates generally have a larger brain size for their bodies than other mammals. Humans, as it turns out, have unusually large brains even for primates. The human brain is the largest primate brain, measured in absolute terms or corrected for body size, and is significantly greater in size than would be predicted from the allometric relationships alone. Even during the span of human evolution, the brain has increased relative to body size, with a consistent increase from *Australopithecus*, *Homo habilis*, *Homo erectus*, through *Homo sapiens*. This at least suggests selection for increased brain size, even when corrected for body size as required.[18] The meaning of such comparisons is controversial. It is not clear how encephalization quotients (the ratio of cortical area relative to body size) or how allometrically adjusted brain size would explain differences in ability. Realistically, these are still gross measures, which gloss over the kinds of specializations that are likely to explain major differences between species. They also are not sensitive to differences among specific areas, though that could be corrected for. The comparisons among primates are less problematic than are comparisons across broader taxonomic groups, largely because the allometric variables are more constant among more closely related groups. Structurally, primate brains are very similar. They have the same gross cortical and subcortical structures, with a similar architecture. However, the similarity is not perfect. Aside from its larger size, the human brain is more asymmetrical and its olfactory centers are reduced.

None of this goes any significant distance toward explaining the marked cognitive differences between humans and other primates (though, of course, our ability to smell is reduced), such as, most obviously, our specialized linguistic abilities. Yet it seems relatively uncontroversial to conclude that there has been selection favoring increased size, and presumably *for* increased intelligence. Moreover, over the past three million years, estimates of hominid brain size show overlapping distributions. *Homo habilis* had an average brain size, smaller than *Homo erectus* and larger than *Australopithecines*. However, the lower range of H. *erectus* fossils have brain sizes that overlap with the high end of H. *habilis*. The overlap is similar among other groups. This suggests that there have been no sudden transitions, but rather a continuous evolution toward larger relative and absolute brain size. This kind of result does lend some credibility to the general claim that there has been selection for increased intelligence within hominid lines. It would be consistent with Nozick's conjecture as well as Dennett's. It offers no specific support at all for the claim characteristic of Cosmides and Tooby that there has been selection for some component of human rationality, specialized for social exchange, though it is consistent with that view as well.

(2) *Ecological factors* What ecological basis do we have for selection for any aspects of human intelligence? Consider first Nozick's conjecture that rationality allows us to deal effectively with new and changing situations. The view is not without plausibility. It has long been known that in spatially or temporally variable environments, there is an advantage to variation, and when the environmental variation is relatively fine-grained—that is, when an individual is likely in its lifetime to encounter more than one habitat—selection will favor response plasticity (see Levins 1963 and 1968; Lewontin 1957; and for experimental confirmation, Sultan and Bazzaz 1993a, 1993b, 1993c; Day, Pritchard, and Schluter 1994). Nozick's conjecture is plausible, but it comes to little more than the common recognition that generalists—who encounter wide variations in environment—will have improved fitness insofar as they are able to accommodate to variation. Humans are certainly generalists. Recognizing this tells us little about how environmental variation is recognized or responded to.

Alternatively, Cosmides and Tooby's suggestion is that human reasoning is the product of selection based on social exchange, which places the emphasis on the social environment. I assume that the social environment is crucial to many social adaptations; but our knowledge of the social behavior of early hominids is, unfortunately, sparse. Neither the behavior in contemporary non-agricultural societies nor that in primate social groups provides sufficient information to know how social structure would have shaped the behavior of early humans or their ancestors. Cosmides and Tooby assume that reciprocal altruism was crucial in shaping human rationality. Once again, this is quite likely true. There is no doubt that the social environment was an important factor in shaping human reasoning. Reciprocal altruism was undoubtedly important in shaping the capacities of dolphins, birds, bees, and ants as well. The constraint is real, and general, but it explains nothing specific to human reasoning. To give it substance, we would need to know much more about the social organization of early hominid groups. We would need to know something about the scarcity or abundance of resources and the social structure of early hominid groups. Were they large or small? What was their organization? Were they loosely structured? What dominance patterns did they exhibit? All of these questions, and more, are relevant to an explanation of social traits of early hominids. What we lack is sufficient information to offer more than the vague recognition that reciprocity was important or that cognition is geared toward variation.

(3) *Heritability*   What can be said about the heritability of human capacities for reasoning? The obvious difficulty is that we do not know what, if any, genetic variation is relevant to the question. Almost three quarters of human genes are monomorphic, that is, showing no variation within or between populations. Of the remaining variation, the bulk is among individuals within groups (see Lewontin 1982). In ancestral populations, we have nothing to go on. We also have no idea what would constitute a reasonable measure of intelligence in interspecific cases. Our most common measures of human intelligence are largely linguistic and would be no more satisfactory as measures of relative intelligence than would differences in echolocation or olfactory capabilities. We can make little sense of interspecific comparisons of intelligence, and correspondingly little sense of the necessary claims concerning its heritability. Cosmides and Tooby are reasonable and clear in their own assessment of the situation:

We assume that the cognitive programs of different individuals of a species are essentially the same—that cognitive programs are species-typical traits. However, the parameters fed into them can be expected to differ with cognitive circumstance. . . . Cognitive programs constitute the level of invariance for a science of human behavior, not behavior itself. (1987, 284)

Unfortunately, this leaves us with nothing revealing to say about the heritability of these cognitive programs. If the "cognitive programs" are the same, then there is no variance, and so we cannot bring the apparatus of contemporary population genetics to bear at all. If there are differences, then it can be brought to bear, but we would need measures of these differences. We have no such measures. So we are left with nothing.[19]

(4) *Population structure*   What do we know of population and environmental structure? Again, the short answer is the right one: very little or nothing. We do know that the evolution of hominids was accompanied by substantial environmental changes. We do not know how that would have affected the breeding structure of populations. We do not know what the relevant population size was, or the degree of gene flow between groups. We do not know the factors that would likely have been most significant in dealing with human cognitive evolution.

(5) *Trait polarity*   Finally, it is at least easy to see that human rationality is a derived trait. We have no information whatsoever concerning the ancestral traits. We do not know what preconditions there were or how they were modified. So although we can be sure that human rationality is a derived trait—if for no other reason than that there has been an increase in brain size—and although we know our ancestors had smaller brains than we do (on average), we unfortunately do not know in any specific way how these increases were realized.

Again, the overall assessment is less than encouraging. It is reasonable to think there was selection for increased brain size, but the significance of that for the evolution of human reasoning is uncertain. We are sorely lacking in any knowledge of the kind or extent of variation in ancestral human populations. We do not know what kind of selection pressures shaped the evolution of human rationality, or even such critical facts as the social structure of early hominid groups, which would affect both the kinds of selection pressures and the distribution of traits. We are offered various suggestions concerning the evolution of rationality but do not have enough information to resolve the differences between them. Beyond the bare assertion that there has been selection for increased intelligence, the evolutionary record offers us little help.

6   Thinking of Matter

We began with broad speculations characteristic of evolutionary psychology. Human cognition and language are adaptations. The evolutionary function of language is communication, or is at least somehow tied to communication. The evolutionary function of human cognition is, on one account, to enable us to deal with variable and novel environments, and, on another account, to facilitate social interaction. It is remarkable that in neither case does the account of the evolutionary function aid in explaining the features that make it plausible that these are adaptations. Assuming that complexity is symptomatic of adaptation, this complexity should be explained adaptively. In the case of language, it is the character of human language—as a complex capacity, governed at a number of levels by complex mechanisms and rules—that makes language unique and makes it plausible that it is an adaptation. The proposals concerning its evolutionary function, however, reflect none of the complexity; Pinker and Bloom's proposed evolutionary scenario fits ants and birds as well as humans. It sheds no light on the specifically human features that so fascinate them.[20] Nozick's proposal is similarly mute concerning the specific features of human reasoning and, in particular, the features that, in his view, would be necessary to explain. Cosmides and Tooby's proposal at least has the virtue of being relevant to the specific structure they find in human reasoning. If human reasoning is, as they say, designed to thwart cheating, then it is at least clear what conditions would need to be met in order to explain human reasoning. The problem in this case is that the evolutionary information we actually have has little or no relevance to the specific models they offer of human reasoning.

It should be emphasized that the standards I have adopted for what constitutes a good explanation of evolutionary adaptation are restrictive and difficult to meet satisfactorily. Evolutionary biology is not easy. I've shown, nonetheless, that these are reasonable standards, given work within evolutionary biology. The standards more typically embraced within the collection of views that is called "evolutionary psychology" are much less limiting. The approach is one that depends on what Tooby and Cosmides call "functional analysis." It begins with an analysis of what constitutes successful performance of a trait, relative to the presumed ancestral conditions in which it evolved. The central question is then taken to be whether some proposed design would, in those circumstances, have proven to be adaptive. My question is not whether some design might have been adaptive, or even would have been adaptive, but whether the right explanation for the presence of the proposed design, for example, human language and reasoning, is that it *is* an adaptation and how it should be explained. Even given that human language or reasoning is an adaptation, the point I have pressed is that we should not think we have explained the proposed design with the sorts of general suggestions that are warranted. Anything more is raw speculation. When Darwin was confronted with speculations concerning the origin of life, he wrote, in a letter to Hooker, that "It is mere rubbish thinking, at present, of [the] origin of life; one might as well think of [the] origin of matter" (Burkhard 2001, 11:278).

Confusing historical function with contemporary function is one of the failings of human sociobiology. Stephen Gould wrote in a review of Charles Lumsden and E. O. Wilson:

Historical origin and current function are different properties of biological traits. This distinction sets an important general principle in evolutionary theory. Features evolved for one reason can always, by virtue of their structure, perform other functions as well. Sometimes the principle is of minor importance, for the directly selected function may overwhelm any side consequence. But the opposite must be true for the brain. Here, surely, the side consequences must overwhelm the original reasons—for there are so vastly more consequences (surely by orders of magnitude) than original purposes. (1987, 122)

We do not need to follow Gould in thinking that selection will have a diminished role in the evolution of human language or reasoning; in fact, I think we should not. All of the speculations I have canvassed have the virtue of respecting the difference between historical origin and current function, which Gould insists on. In this we should follow Gould. Cosmides and Tooby, for example, emphasize the importance of understanding evolution in terms of the ancestral environment rather than current function. In practice, they evoke some standard models in evolutionary biology, such as the models for reciprocal altruism and kin selection. Assuming that these govern the evolution of traits for social cooperation, Cosmides and Tooby conclude:

the selection pressures analyzed in optimal foraging theory are one component of a task analysis . . . of the adaptive problem of foraging. It defines the nature of the problem to be solved and thereby specifies constraints that any mechanism that evolved to solve this problem can be expected to satisfy. In this case, optimal foraging theory suggests (a) that we should have content-specific information-processing mechanisms governing foraging and sharing, and (b) these mechanisms should be sensitive to information regarding variance in foraging success, causing us to prefer one set of sharing rules for high-variance items and another for low-variance items. (1992, 213)

The important point for our purposes is methodological: Cosmides and Tooby are content to begin with very general models for reciprocal altruism, and then project these as adaptive constraints on the evolutionary "problem" confronting our ancestors. It is, of course, plausible that our ancestors engaged in sharing of food and other resources—so, for that matter, do chimpanzees— and it may be true that there was considerable variance in food availability. But even if true, these general facts and constraints are inadequate to capture the kind of historical information we would need to construct an evolutionary explanation, including factors such as the kind and extent of variation present in our ancestors, the kind and extent of variation present in the ancestral environment, the actual environmental features that affect survival and reproduction, and demographic factors. Whether, for example, we "should" have content-specific information-processing mechanisms governing foraging and sharing is an open question. If, for example, no such mechanisms were part of the natural variation in our ancestral repertoire, they could not evolve. If the preadaptations in our hominid ancestors favored sharing, but the "mechanisms" that mediated sharing were not domain-specific, then these mecha-nisms would have been favored. Similarly, whether these pressures should yield some mechanisms for high-variance items and others for low-variance items is an open question. If, for example, the ancestral environment encom-passed high degrees of resource variance, then mechanisms for that contin-gency would be favored, whatever might follow.

I offer no alternative evolutionary scenarios. Rather, the point is that however good the formal models of reciprocal altruism may be, we lack the critical kinds of information concerning the historical contingencies shaping our history.[21] Evolutionary history is the substance of adaptation. Without embracing the role of history, we might as well be wondering about the origin of matter. In the end, that too is an evolutionary question, though not one for evolutionary biology.

4 Recovering Evolutionary History

1 Mental Organs and Human Reasoning

In *How the Mind Works*, Steven Pinker embraces the program of evolutionary psychology with characteristic enthusiasm, proclaiming:

The mind is a system of organs of computation, designed by natural selection to solve the kinds of problems our ancestors faced in their foraging way of life, in particular, understanding and outmaneuvering objects, animals, plants, and other people. (1997, 21)

On Pinker's view, like that of Tooby and Cosmides (see, e.g., Cosmides and Tooby 1987, 294–302), the mind is modular in structure, embodying a variety of specialized "organs of computation," with each module specialized and adapted to some aspect of our ancestors' lives. The structure of each module is, moreover, on Pinker's view, "specified by our genetic program" (Pinker 1997, 21). As adaptations, the operations of these modules were molded to fit the "problems of hunting and gathering" that defined our ancestral environment. The modular structures and the modular functions are selected for. Pinker suggests that the key to understanding human intelligence, unlike many other organisms, lies in understanding its flexibility—that is, the human ability to modify behavior to fit variable and novel circumstances.[1] In a similar vein, as we've seen, Robert Nozick (1993, 113) claims that natural selection should favor capacities whose exercise has increased fitness as a consequence of their reliability. In a locally stable and predictable environment, simple reflexes might suffice. Humans, however, respond to variations in circumstance, and this, Pinker and Nozick agree, is the key to understanding the evolutionary function of human reasoning (see Nozick 1993, 120).

Cosmides and Tooby do not agree. The implication of Pinker's suggestion is that human reasoning is a more plastic, general-purpose capacity, something that Cosmides and Tooby emphatically deny. Instead, they see human reasoning as specialized. They see human reason as, specifically, an adaptation for social exchange, somehow reflecting the social structure of primitive human social groups, whereas Pinker offers a less specific portrait, derived from Tooby and Irven DeVore (1987). According to Pinker, the "cognitive niche" occupied by humans is one in which human intelligence is an adaptation shaped by a variety of features, including our gregariousness, the need for "information," tool use, hunting, and social coordination (see Pinker 1997, 188ff.). Human reason serves the purpose not just of social exchange, but of many other purposes as well. Nozick is interested, similarly, in articulating the "evolutionary function" of rationality, which in turn depends on the processes that historically shaped and maintained it. Nozick is studiously vague about precisely what this function might be beyond the idea that rationality enables humans to cope with "new and changing situations."

Nevertheless, the thought that reasoning serves many purposes is not alien to Cosmides and Tooby. Reasoning undeniably has many uses, whatever its proper function might be. They ask at one point why our "ancestral hominid foragers" evolved mechanisms responsive to other hominids; or, as they say, how they came to have mechanisms geared to "reconstruct the representations present in the minds of those around them" (Cosmides and Tooby 1992, 119). Their answer is not much different from that of Pinker or Nozick. Individual experience is a difficult teacher; knowing what is good to do in local conditions by trial and error involves considerable error. Error is costly and often fatal. Relying on others who have experience of what is useful and what is not is itself useful. So reliance on "inferential reconstruction becomes common enough in a group, and some representations begin to be stably re-created in sequential chains of individuals across generations" (ibid.).

In *Darwin's Dangerous Idea*, D. C. Dennett, similarly, defends the centrality of biological adaptation in understanding human reasoning. In defending the thought that "biology *is* engineering" (1995, chap. 8), Dennett proclaims that "Darwin's great insight was that all the designs in the biosphere could be the products of a process that was as patient as it was mindless, an 'automatic' and gradual lifter in Design Space" (ibid., 188). Dennett says that as we move among lineages, we find there are some "forced" moves and some "good tricks." The idea is clearly that, as in chess, some moves must be made and others, if not forced, are just good moves. Dennett's thought is that in the evolutionary "game," the successful ones must "grope toward the Good Tricks in Design Space" (ibid., 307). The idea that there are attractors in design space that can be used to explain evolutionary convergence is important.

Griffiths (1996) notices at least three problems with this thought as Dennett describes it: first, there is a lack of biological examples in Dennett's discussion; second, if the thought is that similarity must be explained by the advantages that some design offers, independent of common descent, then that neglects the crucial historical context that gives substance to evolution; and, finally, showing that a feature is convergent does not show why it is so. There are many biological examples of convergence, some of which I will discuss later. I will also, eventually, have much to say about whether and when similarity is a function of descent or modification, and when it is not. More importantly, Griffiths says that there are numerous problems with adaptation as a methodological approach. Symons (1992), as I've already said, suggests that male sexual attraction is designed to reflect the fertility of postpubertal females. Men who like younger women are supposed to have greater reproductive output, in the long run. Griffiths notices that if the "empirical" results had been otherwise, there would have been an equally defensible adaptationist scenario: if men preferred older women, that could be "explained" by pointing out that older women are proven as mothers, and so males were preferentially attracted to more "mature" females. In many ways, this is *the* challenge of adaptive thinking. There is an adaptive scenario fit for virtually any empirical possibility. It will be at the center of my focus in this chapter.

As always, Dennett's imagery is striking, which no doubt contributes to making it attractive. Without denying that there are differences between the cognitive endowments of humans and other animals, and without denying the importance of culture, Dennett recognizes that "the brains we are born with have features lacking in other brains, features that have evolved under selection pressure over the last six million years or so" (1995, 371). These brains were improved over tens of thousands of years, gradually improving, within a "design space" reflecting efficiency of design. The critical recognition, according to Dennett, is that anticipation has its advantages over trial and error from a Darwinian perspective, just as learned responses have advantages over fixed responses from a Darwinian perspective. It is at least true that in some contexts, anticipation and learning have advantages over their competitors. We need to assume at least that our environment is locally constant, so that recent experiences will predict future ones. Often, that is a reasonable and reliable expectation. The use of information—typically information derived not from our ancestors, but from our contemporaries—gives us a systematic advantage because these contemporaries have experience of the local environment; and Dennett observes that that advantage has allowed humans to spread broadly across the earth, whereas ancestors apparently less adept did not.

The convergence of Dennett's view with those of Pinker, Nozick, Tooby, and Cosmides is transparent. The conditions under which anticipation is better than trial and error, or under which learning is better than fixed responses, are the kinds of conditions under which flexibility would be favored. In an environment that is relatively uniform, in which a local sample is representative, the efficiency of fixed responses would favor them, and there would be no need for flexibility. Once evolved, this flexibility would facilitate our spread to diverse environments. One key to the novel use of information characteristic of humans, Dennett thinks, is language. Neuropsychology has revealed structures in human brains that are unique and implicated somehow in language use. He says,

There is no question that the origin of language is theoretically a much easier problem than the origin of life; we have such a rich catalogue of not-so-raw materials with which to build an answer. We may never be able to confirm the details, but if that is so this will not be a mystery but only a bit of irreparable ignorance. (Dennett 1995, 389)

Dennett likely underestimates how difficult the problems are over the origin of life (see Orgel 2004 and Shapiro 2006 for overviews). My view is that it is an exceedingly difficult question, suffering from considerable lack of evidence concerning the initial conditions. It may yet yield to evidence. Dennett also likely underestimates how difficult the question of the origin of language is. Nonetheless, he may be right that the latter is easier than the former. Dennett is largely following Pinker and Bloom, who earlier defended the idea that human language is the product of Darwinian natural selection, designed to facilitate communication and coordination. Dennett doesn't tell us much about the "not-so-raw" materials we have to work with in this case. Pinker and Bloom (1990, sec. 3; 1992, 454ff.) assume that adaptive complexity requires an explanation in terms of natural selection. The general line should be familiar by now: Human language involves a complex capacity, governed by rules and representations that are not taught in any straightforward way; there are rules governing phrase structure that provide information about the underlying structure and meaning; grammars depend on lexical and phrasal categories that simultaneously limit what can be said and give language its remarkable expressive power; and more. Pinker and Bloom (1992, 459) exploit these general features in concluding that language shows "signs of design." And again, language does seem to be a natural candidate for explanation in terms of evolution by natural selection. It is a remarkably complex faculty. The question is exactly what explanation we should offer. If Dennett is right that the details may forever elude us and that we may well be faced with an "irreparable ignorance," then whether the problem is the origin of life or the origin of human language, an answer advanced in the absence of knowledge of the details is a hypothesis without foundation. It is but speculation.

There is nothing inherently wrong with speculation, so long as it is not confused with knowledge. We would not accept an explanation of the origin of life without knowing the details. We would not accept an explanation of the flower structure of orchids without knowing the details. We should not accept an explanation of the origin of human language or human reasoning in ignorance of the crucial details. An answer that is offered in spite of an "irreparable ignorance" of the details is no more than unbridled speculation. It does not warrant our assent.

2 Adaptation and Evolutionary History

An *adaptation* is a trait that is present or was maintained because of the selective advantage it offered to ancestors; in this sense, to claim that something is an adaptation is to make a historical claim (see Brandon 1978; Burian 1983; Sober 1984; Griffiths 1996). This use of the term is common, but it is not the term that matters. What does matter is that adaptation as a form of historical explanation features importantly in evolutionary explanations. To claim that some trait or behavior is an adaptation involves a claim not about present function, but about the past. More specifically, to claim that some trait or behavior is an adaptation requires that the trait or behavior is present because, in the past, it made a difference to survival or reproduction among ancestors. An adaptation must, as Richard Burian says, be a "consequence of . . . design differences" (1983, 307). This has a variety of implications for an evolutionary theorist. Take any trait at all, from the beaks of finches to eyes of cave fish to the wing shape of *Archaeopteryx*. If these are adaptations, this requires that, in the past, there were differences in this trait among individuals. Otherwise there could be no selection. That does not mean there must still be such differences, only that there must have been such differences in the distant past, differences that must have mattered for survival or for reproduction. Otherwise, they would not have mattered in a way that mattered biologically. Additionally, to claim that some trait or behavior is an adaptation requires that the trait or behavior be *heritable*. If they are not passed on, they cannot matter for evolution. Darwin assumed that most traits were inherited, and by that he meant they were passed on genetically. This claim can mislead. Evolution does not require that the trait or behavior is under genetic "control." It does require that there be a correlation between parent and offspring that makes the trait heritable. Together, these conditions require that adaptations are the consequence of *heritable variation in fitness* (Lewontin 1970; see also Brandon 1978). This is evolution by natural selection: this is Darwin's gift.

Evolutionary psychologists not only recognize but embrace the importance of history and its centrality to evolutionary explanation. It is no part of their argument that our psychological faculties are adaptations simply because they are adaptive in our current circumstances; indeed, our faculties may not be adaptive in our current setting. What is crucial for the evolutionary claims at the heart of evolutionary psychology is that our psychological faculties were adaptive in our ancestral environment, or what is often, but rather misleadingly, described as the "Pleistocene environment." Their *current* presence must be a consequence of this *historical* adaptedness. The claim of evolutionary psychology thus is *explicitly* historical. This is one dimension, I've suggested, in which they have at least improved over the claims of human sociobiology. This raises the central challenge that I've focused on, one that is certainly not insurmountable *in principle*: we cannot, except in unusual and limited circumstances, directly observe the action of selection. Instead, we must infer what processes were present on the basis of the products we observe. We must infer the historical process based on currently available variations and the fossil record. This is, as it turns out, not a trivial limitation, even though it is one that concerns practical limitations.

Here is what Tooby and Cosmides say in defining what it is to be an adaptation:

Stripped of complications and qualifications, an adaptation is (1) a system of inherited and reliably developing properties that recurs among members of a species that (2) became incorporated into the species' standard design because during the period of their incorporation, (3) they were coordinated with a set of statistically recurrent structural properties outside the adaptation (either in the environment or in the other parts of the organism), (4) in such a way that the causal interaction of the two (in the context of the rest of the properties of the organism) produced functional outcomes that were ultimately tributary to propagation with sufficient frequency (i.e., it solved an adaptive problem for the organism). (1992, 61–62)

I am often unsure exactly what they intend. But more "complications and qualifications" would be unlikely to help. We can work backwards to gain some foothold. The fourth condition seems to require that traits which are adaptations must enhance fitness, at least in the ancestral conditions. The third suggests a recognition that traits have specific fitness values only in a given environment, and perhaps that the relevant environment is the ancestral environment. The second indicates that it is the enhanced fitness that caused the traits to become "incorporated into the species' standard design." Finally, the first requires that adaptations be inherited and be reasonably stable features of the organism. So, as Cosmides and Tooby apparently understand them, adaptations are stable and recurring features characteristic of a species that are

stable and recurring because of selective advantages they offered to our forebears in the ancestral environment. Alternatively, they offer this:

Adaptations are mechanisms or systems of properties crafted by natural selection to
solve the specific problems posed by the regularities of the physical, chemical, developmental, ecological, demographic, social and informational environments encountered by ancestral populations during the course of a species' or populations' evolution.
(Cosmides and Tooby 1992, 62)

Importantly, Tooby and Cosmides suggest here that adaptations are "species
typical" features common to individuals across both space and time. This is
an unusual restriction, biologically. It is surely true that many adaptations are
not thus "species typical" or universal. There are, after all, differences between
human groups that apparently are adaptive, but where there is no question
of their being "species typical." For example, humans differ considerably in
superficial traits such as skin color, and in features of morphology. Some of
these are likely to be adaptations. Lighter skin enhances the production of
vitamin D; darker skin provides better protection against the sun. Moreover,
not all such variations are only skin deep—tolerance for lactose, for example,
differs considerably among human groups. Native Americans are more likely
to be lactose intolerant, whereas Europeans tend to be much more tolerant.
Tolerance is likely a relatively recent acquisition, in evolutionary terms,
following on the emergence of agriculture. Some variations in hemoglobin
structure—most famously, the variations relevant to hemoglobin S—are clear
examples of adaptations to local selection pressures. They are not species
specific; they are local adaptations, driven for example by exposure to malaria.
Perhaps the thought that Tooby and Cosmides want to entertain is not that
adaptations are species-typical, since that is false; perhaps the thought is that
species-typical features are adaptations. That may be more defensible. Nevertheless, even it is not always true, as the case of skull sutures suffices to show.
Another simple counterexample is the pentadactyl hand, which is more general
and is evidently inherited independently of advantages of that trait to the
species, whatever those might be.

We've already seen that reverse engineering offers little promise for evolutionary psychology when what is required is to explain psychological traits
as adaptations. The core problem is that psychological capacities, such as
human reasoning, are not subject to the kind of constraints that can be used
to construct an evolutionary explanation. The constraints are many, and difficult to specify beforehand. As a result, reverse engineering offers little more
than storytelling. We've also seen, focusing specifically on the case of human
language, that the resources of population genetics offer evolutionary psychology little comfort. Although population genetics certainly can explain the

origin of complex traits, and although there is a substantial and intriguing literature involving the testing of evolutionary claims within population genetics, that too is largely unavailable to evolutionary psychology. Fruit flies have genes whose function is to detoxify alcohol, which is not especially surprising, given their food sources. Fruit flies are a well-adapted subject for evolutionary studies. Likewise, as I illustrated in the previous chapter, the morphology of *Gammarus minus* is one that is amenable to evolutionary analysis. But psychology, in contrast, is poorly adapted as an evolutionary science.

There is a third approach for evaluating evolutionary claims. It is now the method of choice among evolutionary theorists. I've already deployed it implicitly in my discussion of *Archaeopteryx*. It is often called the "comparative method" and depends crucially on the use of phylogenetic histories. This chapter will ask whether the comparative method offers much promise for the purposes of evolutionary psychology. I have no doubt that it is a valuable resource for evolutionary biology. Some of the cases I'll recount show that it is. I also have no doubt that it is a valuable resource for understanding human evolution. In fact, much of what I say will support that thought too. However, it is another thing to ask whether the comparative method will fulfil the needs of evolutionary psychology. The verdict there will be negative.

3   The Comparative Method

What is often called the "comparative method" has recently gained currency among evolutionary biologists in studying macroevolution. It is equally applicable in microevolutionary contexts. To propose that a trait, behavior, or character is *adaptive* entails that that trait confers an advantage relative to other organisms that lack that trait. To propose that a trait, behavior, or character is an *adaptation* entails that that trait is present because in the past it did confer an advantage relative to other organisms that lacked that trait. Adaptations are thus doubly relative. A trait is adaptive relative to alternatives that are present. Correlatively, an adaptation must have been adaptive relative to historically available alternatives. Moreover, a trait is adaptive relative to a given environment; what is adaptive in one setting may be maladaptive in another. Correlatively, an adaptation is a trait or behavior that was adaptive in some past environment, and thus relative to its previous ecological setting. The comparative method accordingly requires that we compare a trait or behavior to phylogenetically related ancestors and conditions. These constitute the evolutionarily relevant alternatives within related evolutionary lineages. If it turns out that a trait would not enhance performance or survival relative to the actual variants among its ancestors, then that would undercut the claim that that trait is in fact an adaptation. Similarly, if a trait originated in a lineage for which the current function would have been irrelevant, then that would undercut the claim that that trait is in fact an adaptation, whether or not it is currently adaptive.[2] Skull sutures fit in this category. To draw again from Darwin's *Origin*, there are structures—hooks—on the branches of a climbing variant of bamboo. Darwin was not oblivious to the adaptive functions such hooks serve for these climbing species, but nonetheless he rejected them as adaptations. The reason is simply that those same structures are present in other, nonclimbing relatives. Since the structures precede their "function," that function cannot explain their presence.

The importance of using a comparative standard is not difficult to discern. An analogy may help. In work on statistical reasoning, it is imperative to use base rates. Suppose we want to know whether some medicine is effective against some particular disease. It is important to some of us, for example, that aspirin reduces the chances of heart attacks. How do we know that? There are broad statistical studies comparing rates of heart attacks among those who take aspirin with the rates among those who do not. The chances of heart attack in the former are lower. *This* is a comparative matter. In the absence of base rates, we will likely be misled. To use a standard philosophical example from Wesley Salmon, it is true that those who take vitamin C will almost always recover from colds within a matter of weeks. Vitamin C is not a treatment, though, since the chances of recovery are not improved compared to those who do not take Vitamin C. We need to compare recovery rates under various treatments, or in the absence of treatment altogether. It is a difficult matter to know what the appropriate comparison is. If we want to know whether psychotherapy is effective as a treatment for some problem or other, we need to compare recovery under a psychotherapeutic regime with some other group. We know that those who are treated, somehow or other, do better than those who are not treated; so the relevant comparison is not with those who are not treated at all, but with those who have alternative treatments. This is analogous to the rationale for using placebos in drug trials. We compare in these cases, patients who receive a treatment with others receiving a pseudo-treatment.

In evolutionary studies, we face exactly the same problem. We find some trait or behavior in a population. We want to know if it is an adaptation to the environment characteristic of that species. If it turns out that the presence of the trait or behavior is no more likely in the environment that is characteristic for the species than it is for other species, or in other radically different environments, then we should not think the presence of the trait is explained by the environmental conditions. If we recall the situation of *Archaeopteryx*, that should suffice to underscore the point. Dennett claimed that the structure of its foot supports the idea that *Archaeopteryx* was arboreal. If the thought is that the reversed hallux is adaptive for perching, then the point is insufficient. As I observed in chapter 2, *Velociraptor* also had a reversed hallux, as did many of those in the wider group. However amusing it might be to imagine *Velociraptor* perched in the local pear tree where the partridge should be, the foot structure is to be expected independently of the proposed function. The comparative point is decisive against Dennett.

The comparative method allows us to elude the pitfalls noticed by Gould and Lewontin, which they think are common in reverse engineering. Where engineering analyses are silent on history, the comparative method makes evolutionary history the centerpiece of the analysis of adaptation. History is once more given a voice by appeal to phylogenies (see Brooks and McLennan 1991). A phylogeny essentially gives us a descent tree, describing the patterns of relationships among a group of species. Phylogenetic analyses can be used to develop a historical perspective on the pattern of acquisition of traits within lineages, with an eye to inferring answers concerning the adaptive character of traits from the order of appearance of those traits, or to showing the traits are in fact not adaptations. Suppose we have a phylogeny in hand, allowing us to compare an array of related species. Some, we know, are more closely related than others. Some have a specified trait and some do not. The pattern of acquisition or loss of a putative adaptation can then be mapped onto an independently established phylogeny.[3] Given adequate knowledge of the relevant selective regimes—the environments—associated with the ancestral forms, it is then possible to compare alternative characters for their adaptive potential. The method is in fact a powerful one. The method also imposes severe requirements: we need comparative data on related taxa, relevant developmental information, information concerning the character of the environment, the trait family under consideration, and the relative adaptiveness of the traits characteristic of the several taxa (see Baum and Larson 1991; Brooks and McLennan 1991; Coddington 1988; Harvey and Pagel 1991). In many cases, we may not be able to meet the requirements for a comparative study, and even if we do, the method is fallible. Sometimes we don't have the evidence we need. Sometimes, though, we do.

Adaptationist hypotheses have implications for phylogenetic origin and the "utility" of characters relative to alternatives. Such hypotheses are inherently comparative, projecting an advantage for one character or trait *relative to* others; likewise, they depend on identifying one variant form as more fit compared to alternative competing forms in specific environments. Phylogenetic analyses, as it turns out, not only can be used to infer the ancestral conditions, but can also be used to assess the range of ancestral variation and the adaptive significance of a trait. If a character did not provide an advantage relative to variants present at the time it evolved, or if it originated in a lineage for which its currently adaptive function would not be significant, or if the trait is present as often in different environments, then that trait *could not be* an adaptation. The relevant phylogenetic information can naturally be captured in a cladistic framework.[4] The idea is that a cladistic methodology can expose an independent historical constraint on evolutionary explanations. Given a derived trait within a cladistic lineage, it should be possible to identify both the primitive or ancestral traits and their functions. It should also be possible to obtain some idea of the range of relevant variation. This in turn should make it possible to evaluate adaptationist hypotheses.

The basic idea of phylogenetic systematics is straightforward. Given a group of related organisms, it is possible to recover the phylogenetic relationships within that group on the basis of shared similarities in extant species. A phylogenetic tree is typically given as a branching diagram, depicting a series of speciation events; any given phylogenetic tree is essentially a hypothesis concerning the genealogical relationships among a set of taxa. Cladistics is one approach to phylogenetic systematics; it is a straightforward, if controversial, approach. I think it sheds considerable light on the problems. If we begin with a monophyletic group or clade, consisting of an ancestral species and all its descendants, the phylogenetic problem is to determine which of the possible phylogenetic trees is the actual one. With even a small number of species, the number of possible trees can be quite large. To use a simple and artificial example, if we have four species A, B, C, and D, each of the cladograms in figure 4.1 represents a distinct genealogical hypothesis. (These sorts of toy examples are ubiquitous in the literature. I claim no originality. I don't even know where I first encountered this kind of example.) In order to determine which tree is the correct tree, we consider the traits characteristic of the species within the clade. In theory, similar characters could be due to common descent (homologies), or to parallel or convergent evolution (homoplasies). Cladists, following Willi Hennig (1966), recognize that the patterns among shared, derived characters can be used to construct a phylogeny, which in turn should reveal the pattern of common ancestry. They start with the conviction that genealogical influences should be sufficiently pervasive that homologous characters should predominate over homoplasious ones. Convergent and parallel evolution, that is, should be less common than similarity due to common descent. Whenever possible, then, we should assume that similar characters are due to common descent rather than convergent or parallel evolution. The correct tree can then be inferred by examining character similarities, with the

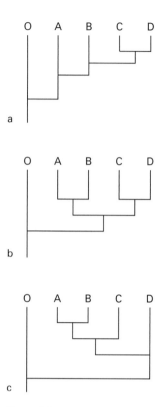

Figure 4.1
Alternative phylogenies. Cladograms describe the structure of monophyletic groups and not, strictly, ancestral relationships. A cladogram groups species in such a way as to reflect evolutionary descent, based on shared derived characters. Trees with distinct topologies represent alternative phylogenies. With four species within a clade (A, B, C, D), each of these three trees represents distinct evolutionary phylogenies. The correct tree, given the character distribution in table 1 and assuming that the most parsimonious tree is one that minimizes character transitions, is figure 1a. The clade consisting of A, B, C, and D is defined by characters 1 and 2, because these are the shared, derived characters within the clade. Characters 3 and 4 in turn define a distinct clade consisting of B, C, and D. Characters 5 and 6 define a clade consisting of C and D. D is distinguished by character 7, and C by the loss of 4. (Within the clade including C, the ancestral state is positive for 4, so it must involve a secondary loss.) Character 8 provides no phylogenetic information, since it is ubiquitous within the clade and the outgroup O.

Table 4.1
An artificial and simplified character set for five species and eight characters

Species	Characters							
	1	2	3	4	5	6	7	8
O	–	–	–	–	–	–	–	+
A	+	+	–	–	–	–	–	+
B	+	+	+	+	–	–	–	+
C	+	+	+	–	+	+	–	+
D	+	+	+	+	+	+	+	+

Assuming discrete characters, a + indicates the presence of the character within the species, and a – indicates its absence. O is the outgroup to the clade.

best tree minimizing homoplasies. To follow our artificial example, suppose we have character distributions among our species for eight traits as in table 4.1. For simplicity, we assume that the characters at issue are binary: a species either has or lacks the character, and possession is not a matter of degree.[5] If we assume that there are as few character transitions as possible—that is, we assume that homology is the rule and not the exception—then it is possible to determine which tree constitutes the best representation of the actual history of the clade. Based on the information concerning the five toy "species" in table 4.1, it is possible to see that characters 1 and 2 define the clade and distinguish it from the outgroup O. Characters 3 and 4 identify a monophyletic group (consisting of B, C, and D). Characters 5 and 6 define a third monophyletic group (consisting of C and D); and the loss of character 4 and the acquisition of character 7 distinguish the species within this group. This is just a matter of grouping according to similarity.

Systematics certainly has had its share of controversies (see, e.g., Hull 1988). Cladism is a controversial method, and even among cladists, there are numerous currently unresolved issues within phylogenetic systematics. These often depend on how to establish character polarities (for a good discussion, see Maddison, Donaghue, and Maddison 1984), and on what counts as a single character. First, consider the question of character polarities. A phylogenetic analysis depends critically on defining an appropriate outgroup, that is, a closely related species outside the clade used for comparative purposes.[6] This outgroup determines the character polarities since it defines the ancestral condition. If a group of bears is white, and its distant ancestors were brown, then white is derived within the lineage and needs explaining, perhaps in adaptive terms. If a group of bears is white, and all the ancestors were white, then white is not derived, and it needs a different sort of explanation. In our simplified example, the stipulated outgroup O has only one of the characters in common with the clade. The only characteristics that provide useful information for an evolutionary phylogeny are those not shared (by descent) with the outgroup since only these give information concerning the topology of the tree. In our example, character 8 provides no phylogenetic information at all since it is ancestral and not derived.[7] The issue of selecting an outgroup is critical precisely because the task is one of determining what the ancestral, or *plesiomorphic*, character state is; the choice of ancestral traits is critical for the construction of the phylogeny. The insistence on using monophyletic groups is, however, widely accepted even if the particular means for determining relevant outgroups is not.[8] The net result, if the data suffice to reach a definitive resolution, is that the phylogeny as constructed can be used to reconstruct patterns of character diversification or patterns of descent; and we'll see that this in turn can be used as a check on adaptationist explanations.

The issue of what counts as a single character concerns the construction of phylogenetic trees. As we reconstruct trees, we assume we have some set of characters, some different and some the same. The cladist assumes that the best tree minimizes change. Suppose we have a correlated change, so that, say, white fur is somehow connected to white claws. We then cannot assume that we have two characters, since the analysis assumes characters are relatively independent. This is a tricky matter, and sometimes it is developmental information that settles the question. This is one key role for considerations of ontogeny and development in evolution.

Assume, then, that there is an independent phylogeny at hand. Adaptations are then those traits, or combinations of traits, correlated with environmental variables, that have arisen within a lineage as an adaptive response to common selective pressures. In evaluating the significance of a putative adaptation, comparative studies require systematic information concerning the ancestral environment and the factors affecting its evolution. These factors may be biotic or abiotic. If the factors are biotic, they may be interspecific, as would be, say, predation. In this case, the organism is involved in an "arms race" (Dawkins 1982). They may also be intraspecific, or social, as would be the case for traits under sexual selection. In evaluating relative performance, one option is to turn to functional morphology, or to engineering design, assessing performance directly. However this is done, though, a comparative study requires assessing the relevant character relative to alternatives, including in particular the alternatives present within the ancestral environment. If the selective regime offers a selective advantage to one variation within the lineage, that is at least consistent with a hypothesis of adaptation.

Comparisons across phyla are useful in evaluating adaptationist hypotheses, and must take into account their respective environmental settings. It's not difficult to see how the reasoning works. Assume the populations or species occupy similar environments. Similar traits across distinct phyla will reflect common ancestry or perhaps stabilizing selection if the traits are homologous. Similarity indicates convergent evolution if the traits are homoplasious, that is, if they arose independently. If the environments are not similar, then similarity of traits suggests that phylogenetic constraints are critical or that there is convergence that is not environmentally driven. Let us return to our toy example. If we observe, say, maternal care in species D and biparental care in C, we may wonder how this difference arose. If maternal care is characteristic of the outgroup and of the other members of the clade, then an explanation of maternal care in D that treats it as an adaptation to changed circumstances of D will be inappropriate. Biparental care is the derived trait within this clade; maternal care is ancestral. There is, minimally, no support for the idea that its presence in D is due to adaptive pressures. The proper explanatory focus would be on the evolutionary origin of biparental care in C. Of course, it may well be that maternal care is an adaptation for some ancestor to the clade. That would in turn require a different comparison, and that maternal care be a derived trait within a wider clade; this does not compromise the point that a focus on maternal care as an adaptation within the clade, or an adaptation specific to D, would be misplaced. It is an ancestral rather than a derived trait. If, on the other hand, we were to observe maternal care in A and D, but biparental care in the other members of the clade and in the outgroup, then that is consistent with the view that maternal care is the adaptation. It is now a derived trait within two lineages, a homoplasy. It needs an evolutionary explanation, and an adaptive one is a real possibility. Coupled with an evaluation of the respective environments of the two species, such comparisons can either support or disconfirm the hypothesis that the trait is an adaptation. An analysis in terms of reverse engineering, by contrast, may tell us a good deal about the adaptive character of maternal care, its current atility, but it will tell us little about whether it is an adaptation, or if it is an adaptation, what it is an adaptation for.

In fact, reverse engineering by itself cannot even guarantee that we are dealing with a derived trait. It is consistent with Fisher's analysis of sex ratios, that a 1 : 1 ratio is ancestral and not derived; it might be, as many suggested early in the twentieth century, that a 1 : 1 ratio among mammals is the simple consequence of Mendelian segregation and a chromosomal determination of sex. If so, then in spite of Fisher's elegant equilibrium explanation, a 1 : 1 ratio may not be an adaptation at all, at least within lineages in which sex is determined chromosomally. Correspondingly, deviations from a 1 : 1 ratio are sometimes derived traits, and those may need an adaptive explanation.[9] What is evidently needed is a return to history, and this is what the comparative method offers.

Three quick examples may serve to illustrate the usefulness of the comparative method for evaluating adaptationist hypotheses, and to suggest some general conclusions that might be drawn from comparative studies. First, some evolutionary scenarios can be definitively rejected using a comparative method. In criticizing the assumption of ubiquitous adaptation, Lewontin (1978) once suggested that the difference between the one- and two-horned rhinoceros is not adaptive. His conjecture was that they are two equally adaptive forms. The two forms might be alternative, but equally good, solutions to an environmental problem, arrived at independently; for example, both might be useful for predator defense, and the difference between them might be insignificant. J. A. Coddington (1988) subsequently pointed out that the simplest phylogeny supports the view that the ancestral form within the clade had two horns rather than none or one and that the Indian rhinoceros lost the frontal horn secondarily. The proper evolutionary question, set within this context, is why this species lost one, but not both, horns. The hypothesis that they arose independently as solutions to a common environmental problem, though, is not defensible given the phylogenetic analysis.[10] The result that these are not phylogenetically independent does not settle the question of whether the loss of one horn was an adaptation, but it does effectively drive home the conclusion that the one-horned variation is an evolutionarily derived trait.

Second, there are comparative studies that offer significant support for adaptive explanations. Anolis lizards are a very diverse group in the Iguanid family (see Losos and Miles 2002). Anoline lizards have proliferated on Cuba, Hispaniola, Jamaica, and Puerto Rico and occupy a diverse range of habitats. On each island, the range of habitats is similar, and the lizards have developed similar morphologies. Some arboreal forms, in particular, have toepads that allow them to hold on to smooth surfaces. (Others that live on the ground lack this particular feature.) Functional studies have confirmed that these toepads are important for their ability to cling, and phylogenetic studies support the view that arboreality evolved before the toepads. Moreover, although there are morphologically similar types on the four islands, similar morphological types have evolved independently; cladistic relatedness is uncorrelated with morphological similarity. So the closest relatives of the arboreal forms are not arboreal. This supports the conclusion that toepads are an adaptation for the arboreal lifestyle of the lizards, and that they arose several times (see Larson and Losos 1996). Generally, the recurrent evolution of similar morphological types in these habitats suggests that adaptation rather than some other process—for example, colonization or developmental constraints—is the cause of the evolutionary changes (see Losos et al. 1998; also Glossip and Losos 1997; Irschick et al. 1996/1997; Jackman et al. 1999).

Finally, the comparative method makes it possible to see what kinds of considerations can make a case more ambiguous. The genus *Montanoa*, so-called daisy trees because of the structure of their florets, includes some thirty taxa. Twenty-one of these species are shrubs, five are trees, and four are vines. The shrub form is apparently ancestral, and the prevalence of this form suggests significant conservatism within the group. Roughly 15 percent of the species are associated with a change in habitat and have given rise to tree and vine forms in roughly equal proportion. A study of convergent adaptation might be appropriate for these forms, following the lines of the previous example; and, of course, an adaptive explanation would be misplaced when applied to the shrub form, which is plesiomorphic. The tree forms are clearly derived, and so are candidates for being adaptations. Among the tree forms, there are numerous morphological changes present, and though they belong to different clades, these characters allow them to live in higher elevations. As far as this goes, adaptation is a plausible hypothesis; we have a derived form, within several lineages, and it is correlated with significant environmental differences. It is in fact not yet clear that an adaptive explanation is right. Diploidy is the ancestral condition among the *Montanoa*, but the trees and only the trees are polyploids. Polyploids are typically larger than diploids within the group, and polyploids often arise from diploids in the group. The distinction between trees and shrubs may in fact be a simple by-product of changes in ploidy level. It may, that is, be a spandrel (for more discussion of the case, see Brooks and McLennan 1991, 159ff.).

The principal moral is that phylogenetic information provides a useful and important independent constraint for framing and evaluating evolutionary explanations. It can be used to rule out some traits as adaptations. It can also lend varying degrees of support to the hypothesis that a trait is an adaptation. The comparative method offers no simple litmus test for adaptation. It may support the claim that a trait is an adaptation, or it may undercut it; but there are a few more specific morals that can be drawn. First of all, unless a trait is derived within a lineage, a comparative method will not support an adaptive explanation within that lineage. (This is, naturally, not equivalent to saying that it undercuts an adaptive explanation altogether.) Second, the claim that a trait is an adaptation is strongest if there is more than one independent evolutionary origin associated with comparable selective regimes, as in the anolis lizards. If a character evolves within a specific selective regime more often than would be expected by chance, then that supports the claim that it is an adaptation within that group. Toepads, for example, have evolved in at least three lineages of lizards. This can be subjected in some cases to a statistical test, evaluating whether an evolutionary transition takes place within a common environment more often than would be expected by chance. Larson and Losos (1996) estimate that the probability of three origins for toepads is roughly 0.028, which supports the claim that it has indeed been selected for within an arboreal habitat. And finally, there is a third moral: since historical variables concerning rates of evolution, ancestral environment, and branching of lineages are more available for some taxa than others, there is no guarantee that a phylogenetic study will yield a definitive answer; nonetheless, adaptation does indicate a number of results that would not generally be expected under alternative evolutionary scenarios.

4 Phylogeny and Adaptation in Human Evolution

The question we now face is whether the comparative method will help the evolutionary psychologist. There is a remarkably good fossil record relevant to the advent of modern humans, and a great deal is known about our primate kin. This might seem a reason to be optimistic about evolutionary psychology. We do, after all, have independent information about the relevant phylogenies based on the fossil record and on contemporary species. There are, moreover, some reasons for viewing human intelligence and human language as adaptations that are largely consonant with comparative methodology. They are, at least, derived traits. However, the grounds for optimism are illusory. On closer inspection, they evaporate. Our psychological capacities are surely the products of evolution. However, despite the rich resources in terms of human evolution, the sort of information that is available does not support the adaptive explanations common within evolutionary psychology. A look at the state of the art in hominid evolution should suffice to convince the skeptical reader. What is known of hominid phylogeny and what is known of the evolution of these traits is just not enough to support the program of evolutionary psychology.

First some additional terminology. The class of hominids (*Hominidae*) is a clade including all those species after we split off from the last common ancestor we shared with apes. It includes all Australopithecus and Homo species. We are in the same clade with the apes, subsumed under the larger family of hominoids (*Hominoidae*). Hominoids are the still broader group of primates, so that apes (which includes humans within the group) form a well-defined clade. There are, unfortunately, only five living genera of apes, and of those only two have more than one living species. The fossil record for the apes is a bit disappointing; there are few fossil remains. The split between hominids and apes is a matter of considerable controversy. The fossil remains we do have shed less light than we would like on crucial evolutionary transitions. The great apes probably originated in the middle Miocene (11–16 mya). The most famous specimen is probably *Proconsul*, though there are several genera representative of ancestral hominoids. (See Ward et al. 1999 for a current discussion. Ward and his collaborators describe a specimen, *Equitorius*, that is primitive but may be a sister taxon. They claim that it is distinct from *Proconsul*, though it shares some skeletal features.) Some apes are more closely related to humans than to other apes, so from a cladistic perspective we need to be included among the apes;[11] we clearly are in the same clade. Hominids evidently diverged from the other apes between four and six million years ago. It is clear that the orangutans (*Pongo*) and gibbons (*Hylobates*) are relatively distant relations of the gorillas (*Gorilla*), common chimpanzees (*Pan troglodytes*), pigmy chimpanzees (*Pan paniscus*), and humans (*Homo sapiens*). The four latter species form an evolutionary clade, though there is less unanimity concerning the pattern of relationship among them—that is, whether, within the hominoid primates, the African apes are more closely related than any other is to their human sister group. The branching events that distinguished gorillas, chimps, and humans were nearly simultaneous; as a result, there is a great deal of similarity among all of them from genetic, morphological, and behavioral standpoints. Morphological evidence tends to favor a grouping of chimpanzees with gorillas, some of which are plausibly construed as sharing derived characters (Kluge 1983). Molecular phylogenies, in contrast, favor a grouping of chimpanzees with humans (Sibley and Ahlquist 1984). Insofar as there is a consensus, it is that the chimps and hominids form a clade, with gorillas branching off first (see Andrews and Martin 1987, fig. 2). One moral from this concerns the choice of outgroup in constructing hominid phylogenies. It seems clear enough that the right choice is chimpanzees, though it is also likely, given the nearly simultaneous split, that gorillas would also be a workable outgroup. In any case, it is our ape friends that should be the outgroup for the purposes of determining ancestral characteristics.

There are, at the current count, perhaps nineteen hominid species, grouped within five or six genera. We are, of course, the only surviving hominid. The oldest hominid species (*Sahelanthropus tchadensis*) was discovered in Chad, and dates to between six and seven million years old. It is in fact very similar to other known apes, though it also has some characteristics of later hominids. Among these hominids, there are a number of fossil forms of australopithecines and several species within the genus *Homo*. All of these are characterized by a bipedal gait.[12] Indeed, a bipedal gait is regarded as one of the key landmarks in human evolution. T. H. Huxley, Darwin's "bulldog," whom we

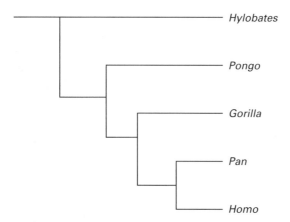

Figure 4.2
A phylogeny for living apes. Though the proper phylogeny is controversial, this represents the most commonly held phylogeny among living apes. It is uncontroversial that humans, chimps, and gorillas are more closely related to one another than to either orangs or gibbons. On the basis of genetic similarity, humans and chimpanzees would form a clade, with gorillas branching off earlier. There are alternative phylogenies, according to which chimpanzees and gorillas form a clade, with the human line branching off earlier.

encountered earlier, emphasized the priority of bipedal gait over cranial capacity. Like other apes, hominids are sexually dimorphic, probably with long gestation periods and extended maternal care. Australopithecines have relatively large teeth with small brains. All the species of *Homo* are broadly distinguished by an enhanced cranial capacity in comparison with the australopithecines. The relationships among these various hominid forms are uncertain and subject to considerable divergence of opinion (cf. Chamberlain and Wood 1987; Strait, Grine, and Moniz. 1997; and Fleagle 1999). The several species of *Homo* do at least form a clade distinct from the australopithecines, though some treat *A. afarensis* as the sister group, while others treat *A. africanus* as the sister group. Among the australopithecines, relations among *A. bosei*, *A. robustus*, and *A. aethiopicus*, and, more recently, *A. garhi*, are especially controversial, though *A. bosei*, *A. aethiopicus*, and *A. robustus* appear to be more closely related to one another than either is to *A. africanus* or *A. afarensis*.[13] One factor that confuses matters is that these species overlap considerably in time, so divergence is difficult to discern. They also overlap with some species of *Homo*. *Homo habilis* diverged from its Australopithecene relatives as much as 2.5 million years ago. There are at least four forms of *Homo* commonly recognized, and sometimes as many as eight (see table 4.2). The principal criterion for inclusion within the genus *Homo* is cranial capacity (greater than 600 cm^3). There are some practical advantages to distinguishing various

Table 4.2
Features of five species within the genus *Homo*

Species	Height	Cranial Capacity	Known Dates
H. habilis	1–1.5 m.	500–800 cc	2.4–1.6 mya
H. georgicus	1.5 m.	600–700 cc	1.8 mya
H. erectus	1.3–1.5 m.	750–1,250 ml.	1.9–0.3 mya
H. neanderthalis	1.5–1.7 m.	1,200–1,750 ml.	130,000–60,000 years
H. sapiens	1.6–1.9 m.	1,200–1,700 ml.	160,000 years

The most marked characteristic of the clade is a larger cranial capacity; many other traits, including a bipedal gait are shared with australopithecines.

species, but for simplicity I will focus on a subset of those often recognized (see Wood and Collard 1999 for an elegant review). Some of these are controversial and fragmentary. It is generally conceded that *H. habilis* is distinct from the remaining groups within the genus. The relationships among *H. erectus, H. rudolfensis, H. sapiens*, and *H. neanderthalis* are less clear; on some accounts the Neanderthals are a subspecies of humans and on others they are a distinct species. Insofar as any distinguishing features are behavioral, the fossil record is not decisive. As a result, hominid phylogenies are very much matters of controversy. Figure 4.3 presents a plausible hominid phylogeny, though not the only one defended in recent literature.[14]

Genetic evidence places Africa as the source of *Homo sapiens*, within roughly the last 160,000 years. Relatively recent finds in Ethiopia that appear to be very similar to modern humans have been dated to between 154,000 and 160,000 years ago (White et al. 2003). The general trends within the hominids do suggest that intelligence has increased. As I've illustrated above, this is essentially a mantra among evolutionary psychologists. Brain size has certainly increased, and the pattern of its evolution—a relatively consistent trend toward larger brain size—supports the conclusion that natural selection was important. Beyond that, caution is necessary. The interpretation of this increase in size is actually exceedingly uncertain. One key problem is that even though size has increased, the functional significance of this is unclear. Size is too simple a variable. As Deacon says, the real question is not just whether size has increased, but "What other changes in brain organization correlate with this global change in brain size, and what are their functional consequences?" (1997, 148).

Still, it is important to understand the patterns we can see. Although the brains of australopithecines fall within the range to be expected for great apes, by roughly 1.5 million years ago, the brain of *Homo erectus* averaged more than twice as large. Gross brain size, as I've said, is by itself unreliable as a

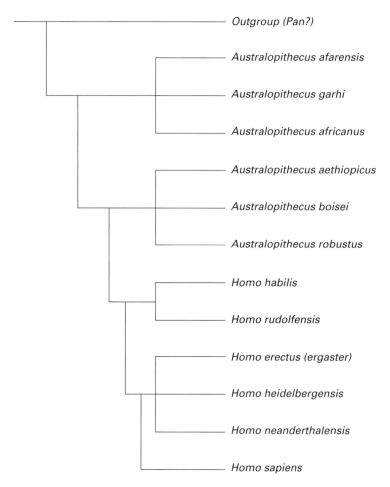

Figure 4.3
A plausible though partial hominid phylogeny. A number of phylogenies have been proposed for the various hominid forms; the resolution of some relationships is not clear. Even the number and rank of differences is contested. Indeed, the number of species is disputed. It is generally agreed that *H. erectus* (*ergaster*) is closely related to *H. sapiens*, that *H. habilis* is more primitive than either, and that *A. robustus, A. boisei* and *A. aethiopicus* are closely related. Some assign these latter three to a distinct genus, though there is controversy over whether they even form a clade. I have depicted this as an unresolved node. There is no clear consensus on whether *A. africanus* or *A. afarensis* is more closely related to the robust forms of the australopithecines. What is given above thus is a plausible phylogeny, though emphatically only one of many defended in recent discussions.

measure of intelligence; it must at least be scaled to body size if we are to make real sense of these trends.[15] Allometric scaling (see Gould 1966; Holloway 1979), which relies on a measure of brain size relative to body size rather than absolute size, further supports the conclusion that the relative cranial capacity of *Homo sapiens* is a derived character and that there was a trend toward greater cranial capacity within the lineage. Within mammals generally, brain size can be reliably predicted from body size as a power function. As a rough rule, brain size scales as a power function of body size within taxonomic groups: for mammals, the scaling factor is approximately 0.75, and for primates the estimates vary somewhat but are more or less consistent with this value. Though there is again some uncertainty concerning such comparisons, it is generally agreed that at any stage of development, primates have a larger brain size for a given body size than would be expected from the mammalian index alone, by roughly a factor of two. (However, mice have a higher ratio than humans. This supports the idea that these comparisons should be limited to closely related species.) Within primates, the growth curve for human infants follows the primate norm until birth; in most primates, the relative growth rate of the body then increases after birth, whereas in humans the brain grows at an elevated rate postnatally. As a result, adult humans also have a brain size that is markedly higher than would be expected given the primate norms (see Deacon 1992a, 1992b). Primates are therefore exceptional within mammals. Humans are unusual even for primates. Such comparisons are suggestive and are consistent with selection affecting brain size, but they must be regarded with caution. The comparisons among primates are less problematic than are comparisons across broader taxonomic groups. In general, comparisons are more secure among more closely related groups. Structurally, primate brains also are very similar. They have the same gross cortical and subcortical structures, with a similar architecture. However, the similarity is not perfect. Aside from its larger size, the human brain is more asymmetrical than are many primate brains, and its olfactory centers are reduced. Endocranial values alone do not reflect such structural differences. None of the differences in overall size, in any case, do anything to explain the marked cognitive differences between humans and other primates, including, most obviously, our specialized linguistic abilities (see Holloway 1983). If we take seriously the idea, favored by many evolutionary psychologists, that our cognitive organization is modular, then overall differences likely would not directly reflect the evolution of human cognitive capacities. Organizational and functional differences *are* surely more important than sheer bulk in assessing human intelligence and our communicative facility, and on this the fossil record is largely silent. It is not clear how even the allometrically corrected measures would explain such differences in ability. Realistically, encephalization quotients (which describe brain size relative to body size) are crude measures of intelligence at best. They reflect overall size, but gloss over the kinds of specializations that are likely to explain major differences among species and that are likely to be important to human cognitive capacities. Nonetheless, this enhanced cranial capacity is at least a derived feature within the genus. That is at least consistent with selection for enhanced intelligence, for whatever that is worth.

Language provides a similar case, though it is actually somewhat more difficult (see Holden 2004). In the last few years, a wide range of work has been done on the origin of speech. Steven Pinker and Paul Bloom (1990) led the charge. Our focus is specifically on the evolution of language. Once again, as I've admitted, there is much that supports the view that human language is an adaptation. The complexity of our vocalizations and the rule-governed character of language are unlike other systems of communication. The facility of first-language acquisition suggests, moreover, that there is significant specialization for language acquisition (see Pinker 1994). Once again, the neurophysiological changes are the clearest. Humans have a larger prefrontal cortex than would be expected. Cortical regions adjacent to the Sylvian fissure are directly involved in language use. There is now increasing evidence that changes in subcortical structures are also important. Studies of patients with trauma, as well as stimulation studies involving these regions, show that they are specialized for language use. Studies that rely on brain imaging with fMRI support the general picture, though they have shown that language functions rely on a diverse range of brain systems, especially in the left hemisphere. Although there are functional and morphological asymmetries between hemispheres in other species as well, the extent of the asymmetry in humans is certainly more marked than in other extant primates. Moreover, other primate calls are controlled in the midbrain and the forebrain rather than in the motor cortex. As we turn to comparisons with other hominids, the picture is less clear, largely because we must rely on indirect evidence. Fossil remains do allow us to discern some of the gross features of hominid brains since the surface features of the brain leave some imprints on the skull. Endocasts of fossil skulls display these features, though, again, changes in size and other changes resulting from upright posture make the interpretation of these features difficult. The Sylvan fissure does show some asymmetry in archaic humans, including both *H. habilis* and *H. erectus* (Holloway 1983). That would support the view that there is an enlarged left hemisphere and an asymmetry similar to that common in humans. Some researchers conclude on the basis of such evidence that these other species of hominid had specializations associated with language, but though such asymmetries are consistent with the conclusion, they are at best equivocal evidence for language use in these species. Yet others, most notably Deacon (1997), claim that the evolution of human language and cognition is ineluctably intertwined. The most reasonable conclusion, in my view, is that we simply do not know whether language use is a derived character specifically for *Homo sapiens*, or whether it is a characteristic of other hominids as well. It is, though, likely to be a derived trait within the hominid line and an adaptation characteristic of *Homo* to the exclusion of the australopithecines.

Thus, it is plausible to hold that language use and human intelligence are derived traits, but this provides only marginal comfort to evolutionary psychology. Our knowledge of the cladistic structure of the relevant species is imperfect, even though it is very good. There is good reason to hold that both human intelligence and human language are apomorphic; that is, they are both derived from and differ from the ancestral conditions. Our primate cousins simply do not have comparable linguistic abilities, and that is doubtless the ancestral condition. The fact that linguistic ability is derived is at least consistent with the claim that they are adaptations. One thing that is strikingly absent is a knowledge of the relevant homologous structures or behaviors in ancestral groups: that the trait is original to *Homo* tells us only what those ancestors could not do; it tells us nothing about the capabilities they did have. In assessing evolutionary adaptation, information about the range of hominid capacities provides a critical feature in understanding the variation, and thus the relative advantage that a structure or behavior provides. It is not clear, as I've said, whether language use is a derived trait within *Homo sapiens*, or a more general feature of the hominids. It does appear that through the lineage there was a tendency toward greater cranial capacity. This at least is not a feature unique to our own species, but this provides little information concerning the evolution of human intelligence and is even less relevant to the sorts of changes necessary for language. This makes a comparative assessment particularly problematic.

The ecological settings for the several species, insofar as we can reconstruct them from paleoecological data, are varied. During the Pliocene and early Pleistocene (roughly 4 to 1 mya), australopithecines were widely distributed in southern and eastern Africa. The hominid tree was then one with many branches. The characteristic habitats included wooded regions, open savanna, and grassland. *Homo habilis* and *Homo erectus* overlapped temporally with several australopithecines, but some suggest the former were evidently associated more with wooded habitats. With *Homo erectus*, we find more use of open habitats, perhaps indicating increased reliance on hunting. *Homo erectus* was a widely varying species, both in terms of morphology and range; fossils have been found in Africa, Asia, and Europe. By roughly one million years ago, toward the middle of the Pleistocene, the range of hominids had spread out into Europe and Asia. It is anybody's guess what physical environment is associated with the trend toward increased cranial capacity. In any case, by the time humans spread out into Asia and Europe, these changes were entrenched. Our intelligence allowed or facilitated the spread across the globe, opportunistically. The ecological information we do have is varied, indicating a wide range of environments, and little of it concerns what ecological associations would be involved in the evolution of increased intelligence or the use of language. *Homo sapiens* are a relatively recent addition to the scene, dating to perhaps as far as 160,000 years before the present. We are left considering traits that may be ancestral to *Homo sapiens*, or may be specialized, with an uncertain ecological connection. It is hardly an ideal case for a comparative analysis and provides little substance to support any particular adaptive explanation.

When we turn to language, the very feature that makes it plausible to treat it as an adaptation—its uniqueness relative to other forms of communication—makes it an especially poor subject for comparative analysis. Deacon (1997) observes that one serious problem with evolutionary accounts for the origin of language is the lack of continuity with other forms of communication. Pinker notes this discontinuity. He emphasizes it, in fact. What he fails to recognize is that it is a problem rather than an asset for evolutionary explanations. The same observation would be appropriate when we turn to human intelligence, especially if it is not a general-purpose mechanism.[16]

Language acquisition provides a useful model for seeing the depth of the problem facing evolutionary psychology. Linguists, following Chomsky, have recognized that the problem with treating language acquisition as a matter of learning—as some sort of social conditioning or generalization—is that the sort of evidence that is available to the child could not result in what we see. The human child develops language of remarkable complexity, with an ease and facility that is also stunning. Linguists often emphasize the remarkable strides children make in learning language at an age at which other cognitive skills are not especially well developed, suggesting that language acquisition is mediated by distinctive mechanisms. Human language acquisition is so unlike what we think of as learning in other contexts that many linguists are reluctant to count the acquisition of language as *learning* at all. The uniqueness of what happens in language acquisition may be overstated somewhat, but not in comparison to the kinds of mechanisms simple models of learning can offer. We lack the kinds of incremental stages that would explain its acquisition in these terms. Accounts of the evolutionary origins of language are in an analogous position, for much the same reason. The very lack of homology between human language and other forms of communication—even with the forms of communication that are present in other primates—leaves us unable to correlate it in any interesting way with any environmental variables, or to compare it with other forms of communication.[17] Social organization is surely involved somehow in the evolutionary origins of language, given the assumption that language is the product of natural selection. Our primate cousins also participate in complex social arrangements, but there is no real analogue of human language in the rest of the extant primates. We do not know what forms of communication were present in ancestral hominid species, or even whether they had linguistic capacities, for the simple reason that the sort of information available in the fossil record is indecisive. The uncertainty concerning the relevant taxonomic level for the emergence of language only exacerbates the problem.

The mere fact that social coordination is facilitated by the use of human language provides no comfort, for though coordination and social interaction favor some form of communication, there is no basis for thinking that they would favor the sorts of complex features that are integral to human language. If there is any generalization concerning primate social organization, it is that it is highly variable. Gibbons are monogamous, orangutans are largely solitary, gorillas are polygamous, and chimpanzees exhibit a fluid and variable social structure. This makes the extrapolation of ancestral social organization highly problematic. There is virtually nothing that would help in understanding the evolution of language. If we were seriously asked how social organization could favor, for example, the particular lexical and phrasal categories of human language, or how the rules underlying meaning and structure could contribute to enhanced fitness, it is hard to imagine how to answer the question. We can assume that selection would favor a common structure, but there is little explanation for this particular structure, or how it would be adaptive. In any case, all the hominids, and even all primates, are social creatures; and little is known concerning the specific social organization of ancestral hominids. As Deacon says:

A full evolutionary account cannot stop with a formal description of what is missing or a scenario of how selection might have favored the evolution of innate grammatical knowledge. It must also provide a functional account of why its particular organization was favored, how incremental and partial versions were also functional, and how structures present in nonhuman brains were modified to provide this ability. (1997, 38)

No doubt tool use, hunting, social coordination, and communication were significant features in the shaping of human capacities. Beyond such vague observations, we are offered little that is concrete in explaining the specific features of human cognition or language use. If the goal is one of explaining the specific organizational patterns of human cognition, such an explanation is lacking in evolutionary psychology. Pinker, Tooby, and Cosmides emphasize the modular character of cognition; but no explanation of its modular structure is forthcoming. Pinker, Bloom, and Dennett are fond of emphasizing the need for communication and extolling the virtues of social coordination; but this suggests no explanation whatsoever of the various features characteristic of human language use or of the remarkable features of human language.

The fossil record of hominid evolution assembled by anthropologists is remarkable for its quality. It leaves no serious doubt that humans are the result of evolution. Though that record sheds a tremendous amount of light on our ancestry, it does not favor the sort of speculations concerning the evolution of human intelligence and human language offered within evolutionary psychology. In the ideal cases for the comparative method, there is a sufficiently diverse taxonomic group in which we can see the same traits evolving in several lineages under similar environmental contingencies. Convergent evolution, resulting in a polyphyletic group displaying analogous features, is suggestive of natural selection, though even convergence can be the result of alternative mechanisms (see Leroi, Rose, and Lauder 1994). Under the best of circumstances, comparative studies may in fact not be able to disentangle the various features affecting evolution, including selection, drift, mutation, and migration. Under the best of circumstances, it is a difficult task to disentangle one evolutionary explanation from another. Hominids and primates simply do not provide the ideal group for comparative psychological and social analysis. We are a small group with few extant species, and the fossil record does not offer the sort of evidence we need. Moreover, much of what we are most interested in is unique to the hominid line. Much as we might care about our origins, much as we might want to understand the origins of sociality, intelligence, or language, the case humans provide for evolutionary biology is not an easy one. The record is simply not sufficient to answer the questions posed. This may be as good as it gets.

5   The Problem with *Homo sapiens*

The comparative method has been a powerful and important biological tool for discerning evolutionary history and for revealing the importance of adaptation. It is particularly powerful when it can draw on patterns of convergent evolution. Species that occupy similar ecological circumstances often have similar morphologies, as the case of cave organisms attests. These similarities can be largely independent of their genealogy. Generally, a comparative method can be supported by the use of an appropriate phylogenetic analysis. In the case of human evolution, there is a rich fossil record, and it is possible to construct reasonable phylogenies. This might give us some hope that we could use a comparative method to ground the hopes of evolutionary psychology.

The problem facing evolutionary psychology is that the evidence available does not allow us to bring the method to bear in understanding human psychology, even if it is enlightening concerning human evolution more generally. Understood as an evolutionary hypothesis, what we are offered by evolutionary psychologists is inadequate. The complexities of human language and cognition suggest they are adaptations, and it is these complex capacities, among others, we wish to explain. What we are offered is often so general that it cannot approach the problem we begin with. Pinker asks, "Why did brains evolve to start with?" and offers a simple and straightforward response: "The answer lies in the value of information, which brains have been designed to process" (Pinker 1997, 175). Dennett similarly touts the virtues of anticipation over trial and error. It is, of course, uncontroversial that brains do serve to process information, and it is uncontroversial that we do sometimes anticipate the consequences of our actions. It is a long way from such simple observations to an evolutionary explanation of human intelligence, much less to the features more specific to human intelligence. It is as if we began by noticing that the knee is a complex mechanism, and accordingly sought its evolutionary explanation. It too is an evolved mechanism, after all. It is a complex mechanism. It would be no explanation at all to be told that the knee is as it is because it is for locomotion, or that there are virtues in being mobile rather than sedentary. It is, of course, true that the knee is involved in locomotion. It is uncontroversial that mobility is important in human evolution. But these generalities leave us uninformed about that peculiar and complex structure we began with. In the case of the human knee, the significant questions concern whether it is involved in locomotion as opposed to climbing. It is with such alternatives that empirical research is crucial. Here the functional analysis matters.

We don't typically see detailed functional analysis within the literature of evolutionary psychologists, though among evolutionary biologists it is common. So in the case of *Archaeopteryx*, a lot of attention has been paid to the strength of the flight stroke. In the case of terrestrial animals, a lot of attention has been paid to the efficiency of gait. In the case of fish, we know a lot about propulsion thorough water. Pinker's and Dennett's specific proposals are similarly uninformative even if they are true. They tell us nothing specific concerning human intelligence. Pinker explains that although there is a common "plan" to the organization of mammalian nervous systems, the human brain is unique in a number of ways. He says:

The human brain . . . tells an evolutionary story. Even a quick side-by-side comparison shows that the primate brain must have been considerably reengineered to end up as a human brain. Our brains are about three times too big for a generic monkey or ape of our body size. The inflation is accomplished by prolonging fetal brain growth for a year after birth. If our bodies grew proportionally during that period, we would be ten feet tall and weigh half a ton. (Pinker 1997, 183)

It is true that our brains are strikingly different from those of our primate ancestors. There is an "evolutionary story" to be told, at least in describing the evolutionary changes within the hominid line. That story does not, by itself, tell us how or why these evolutionary changes came about. There have also been structural changes in the size of the human brain by comparison with our primate cousins or with the mammalian norm. There are, more importantly, changes in its organization. All of this is very indirect evidence concerning the functional differences. It is surely the functional differences that are critical to the ambitions of evolutionary psychology. What explanation is offered for such features uniquely characteristic of the human brain, for example, the reduction of the olfactory lobes? None. What explanation is offered for the features characteristic of human language? Again, none. What Pinker and Dennett do "explain" shifts from the general to the specific, for example, from human language with all its foibles to human language as a system of communication. Systems of communication no doubt facilitate communication. What does that tell us about the evolution of human language? Nothing at all. What does it tell us about the complex features Pinker uses to press for an adaptive story? Nothing at all. It tells us no more about the evolution of human intelligence. This is a theme I noted in the previous chapter. It's a bait and switch maneuver. Initially, we insist on explaining some complex mechanism, noting that if it is complex, it must have been selected for. We then switch to explaining some generic features of the mechanism, ignoring the problem we started with. And then we conclude we were right all along.

Again, considered as an evolutionary hypothesis, the question is what the right explanation of human language or cognition is. The evolutionary psychologist's goal is to provide an evolutionary explanation of human traits as complex *adaptations*—as the products of natural selection—and not merely as *evolved* features of the human lineage. It is important to recognize that there is more than one possible evolutionary explanation even for complex characters, and that not all would make the results adaptations.

It is also important to recognize that even if natural selection plays a significant role, there will still be importantly different explanations possible. Once we recognize the contributions that evolutionary biology offers to evolutionary questions, the fact is that the resources of systematics and comparative biology offer little assistance to evolutionary psychology. Neither do the resources of population genetics. In both cases, the available biological and paleoanthropological record leaves the relevant ecological, social, phylogenetic, and populational variables sufficiently indeterminate that we do not know enough about the factors that shaped human evolution to evaluate the proposals offered by advocates of evolutionary psychology. We do know that language and intelligence are derived forms within the hominids, but it is not clear from the record whether they are features of *Homo sapiens*, or of a broader taxonomic class.[18] If we treat these as specifically human adaptations, for all we know we are focused on the wrong taxonomic level. We do know that social organization and the environment were relevant to the evolution of human language and intelligence, but we know little concerning the actual social structure or even the relevant physical environment. To be told that language facilitates communication is true, but that is hardly sufficient to develop an evolutionary explanation. To be told that human intelligence facilitates the management of information is also no doubt true, but that does not give any substance to an evolutionary explanation of our intellectual capacities. Lacking knowledge of the forms of communication present in our ancestors, lacking knowledge of the sorts of social organization present, lacking knowledge of the sort of ecological problems that these ancestors confronted, lacking even knowledge of which ancestors are the proper focus of investigation, we lack the ingredients for seriously explaining our psychological capacities in evolutionary terms. We are left either with empty generalities or unconstrained speculation.

5 Idle Darwinizing

1 Explaining Descent

Darwin and the brilliant group of naturalists allied with him toward the end of the nineteenth century settled the question of descent. What happened was not just a revolution of Darwin's. Darwin became its centerpiece, its patron saint, but the revolution in thought depended on a network of connections that Darwin carefully cultivated, and who carefully cultivated Darwinism. It was in any case but a decade after the publication of the *Origin of Species* (1859) that evolution came to be broadly accepted among the educated elite. There was of course still resistance to "Darwinism," in some cases by Darwinists. Natural selection did not have the same immediate success, or the same broad acceptance, as did evolution. Even T. H. Huxley was reticent, immediately after the *Origin*, to embrace natural selection, though he eventually came to accept a broader role for Darwin's mechanism. Wallace eventually compromised the importance of natural selection in the case of human consciousness. We now know that the role of adaptation in evolution is a complicated and difficult issue. Darwin and his contemporaries spent considerable energy over the question of whether or not natural selection was sufficient for the evolutionary process. Other mechanisms were also proposed. Darwin himself expressed considerable unhappiness that many missed the subtlety intended in the final sentence of the introduction to the *Origin of Species*, in which he claimed that he is "convinced that Natural Selection has been the main but not exclusive means of modification" (1859, 6). The key thought that was neglected is that natural selection is *not the exclusive means of modification*. The role of natural selection is one that is amenable to a number of different tests.

I have explored three approaches to the evaluation of adaptive explanations, generally more focused on work within the twentieth century than the nineteenth. Each of these three approaches has a significant role to play in evolutionary theory, and I've illustrated their roles with biological cases that have enjoyed broad acceptance and success. Of course, I claim no originality in the research. This is the work of mainstream evolutionary biologists, who work hard and long to establish the cases. I merely report on it. There are many other cases we could call on. These sorts of biological cases provide the impetus for my argument since they set the standards. They define what we should expect of empirically motivated studies of adaptation. More modest empirical cases would not work for orchids, much less for anolis lizards, cave organisms, or even *Archaeopteryx*. Given that evolutionary psychology pretends to drive its reforms *from* the framework of evolutionary biology, those evolutionary standards are the standards that are relevant. If I am right, evolutionary psychology fails to respect the standards we rightly should expect: it fails to respect the standards of the field. As I said early on, if the evidence offered is not evidence of a sort that would be acceptable were we discussing flower structure in orchids, then we should not accept the explanation. Orchids are a paradigm case of adaptation, as Darwin argued; but it is another thing to offer an explanation of what the orchid's structures are adapted for. That is the Darwinian problematic, and it is the problematic for evolutionary psychology. The fundamental problem for evolutionary psychology is that *Homo sapiens* is far less than an ideal evolutionary subject. We are relatively long-lived, and so the more direct experimental studies that attend our knowledge of fruit flies is simply not available, even if we were not ethically constrained. Our knowledge of the relevant social environment in ancestral hominids, which surely did much to shape our cognitive capacities and our emotional profiles, is not something we have much access to from the available record. Even the details of, say, the cognitive capacities of our ancestors is in many ways obscure, though we can see some things in broad outline. The outline is not enough to support any detailed evolutionary explanation for human psychological capacities. We need details that may not be forthcoming.

The three methods I have described are all historical, each one more so than the previous. I began with reverse engineering, one of the centerpieces for many evolutionary psychologists. One of the main problems with the straightforward appeal to design, that of reverse engineering, is that it is not sensitive to changing historical conditions. *Halobates* is an elegantly designed organism, capable of walking on water. What makes a design analysis probative in this case is that we know the physical constraints have not changed over time, and we know what those constraints are. These are physical constraints, and the properties of water have constancy over evolutionary time. Certainly, nothing similar is true concerning *Homo sapiens* and our cognitive and linguistic abilities. We have also seen adaptive thinking illustrated in several cases that are inspired by population biology. These too demand information about the environment that is often not available, including more specifically biological information concerning such things as population structure, the associated environmental factors, and even heritability. The recovery of history that characterizes phylogenetic analyses sometimes gives us just the sort of historical information that is lacking and has been used to recover evolutionary history and propel adaptive explanations to the forefront in many cases. So the anolis lizards studied by Losos and his collaborators provide an elegant case of convergence, which supports adaptation as the cause. The human, and hominid, case offers no similar case for adaptation. Aside from problems already mentioned, the very uniqueness that evolutionary psychologists trumpet undercuts any hope of a comparative analysis. I know of no other empirical methods for evaluating evolutionary explanations of the sorts of traits that attract evolutionary psychologists.

If we are to gain insight into our cognitive evolution, or more generally into the evolution of the mind, we would doubtless need to have independent information concerning the conditions—in this case, the social conditions—relevant to our cognitive evolution. With information concerning the range of variation and the environmental conditions, we often can reconstruct the causes of evolutionary change. Instead, what we are offered are vague suggestions concerning the "Pleistocene" environment. The Pleistocene was named by Darwin's great geological patron Charles Lyell in 1833 and is distinguished by its characteristic fauna. The Pleistocene itself dates to perhaps 1.8 million years ago, after the recession of the seas that were common for most of the Pliocene. In the Pleistocene, we have seven glacial expansions and recessions, including our own current "interglacial interval." One fundamental fact is unassailable: the Pleistocene is a highly variable environment, and so speaking comfortably of "the" Pleistocene environment is a mistake. We are a relatively recent addition, though deeply embedded within the Pleistocene. We diverged from our ape cousins some time in the late Miocene (10 million years ago). The Australopithecines can date to perhaps the beginning of the Pliocene (5 million years ago). Hominids encompass the Pleistocene (with H. *erectus* fossils dating to nearly 2 million years ago). *Homo sapiens* occupy perhaps as much as the last 160,000 years or as little as 100,000 years; very generously, we might span a half a million years. I recount this brief chronology mainly to indicate how loose this talk of the "Pleistocene" is. The Pleistocene covers an enormous amount of time, much more than human evolution, though perhaps convergent with hominid evolution, give or take a bit. It is in any case an epoch with a great deal of change in environmental conditions. It is a mere fragment of "the" Pleistocene that covers human evolution. It is additionally a highly variable fragment, both spatially and temporally. And of course, for adaptationists, the environment is everything. All of this applies to the physical environment, of course. The social environment is even less clear. Doubtless we are social beings. Doubtless, our social conditions affected our evolution. How exactly this is so we just do not know.

2   The Rhetoric of Exclusion

Kitcher observes a key lever used in the rhetoric of sociobiology: "Whoever is not for the program is against Darwin" (1985, 14). This rhetorical technique is equally present in evolutionary psychology. The basic move is evident in Cosmides and Tooby's most aggressive brief for evolutionary psychology. They want us to accept a dichotomy between what they call the "Standard Social Science Model" (SSSM) and the "Integrated Causal Model" (ICM) they favor (see Cosmides and Tooby 1992, 31ff.). The core feature of the SSSM is its commitment to general-purpose cognitive mechanisms, which lack any nativist element whatsoever. The interesting differences between groups are, accordingly, the result of cultural differences and are properly explained as a consequence of strictly social phenomena. Individual differences, as well, are socially conditioned. Even individual differences are cultural products, a consequence of subtle differences in socialization. Tooby and Cosmides' portrayal is very effective. It is also a piece of sophistry, offering a false dichotomy between a manifestly untenable view and their own. The alternative is one that sees no differences between individuals and no biological contribution to individual or social development. I think no serious figure embraces that view since, perhaps, John Watson in the early twentieth century.

Tooby and Cosmides also say that "There is no small irony in the fact that [the] Standard Social Science Model's hostility to adaptationist approaches is often justified through the accusation that adaptationist approaches purportedly attribute important differences between individuals, races and classes to genetic differences. In actuality, adaptationist approaches offer the explanation for why the psychic unity of humankind is genuine and not just an ideological fiction" (1992, 79). There are multiple ironies here, though not the ones Tooby and Cosmides see. One is that Tooby and Cosmides have missed the thrust of the work by Gould and Lewontin, neither of whom deny that there are individual differences. It is, actually, important to their work that individual variation is common. Neither of course denies a role for evolution, though that too is a persistent charge. Both *have* defended the view that whatever differences we find between groups cannot be used to explain differences between individuals, even when those individuals belong to different "groups." This in no way denies that there are differences among individuals within groups. Indeed, to deny there are individual differences would be astonishing. It is crucial to the development of evolutionary biology in the twentieth century that there are differences and that a focus on variation is the key. As Ernst Mayr often insists, to focus on "typological" or "essentialist" thinking is to ignore the engine of evolutionary change. This is as obvious as anything can be in what are called "phenotypic" differences. Some individuals are taller than others. Some are more adept at mathematics. Some are more insightful. Some are more emotionally stable. There is no news here. What then of genetic differences? The thought Lewontin and Gould have insisted upon, often separately, is that we cannot compare individual differences within groups by looking at differences between groups. This presupposes, and does not deny, the existence of individual differences. Are there genetic differences between individuals? Of course. Among biologists, Lewontin earned his stripes—or his stars—by demonstrating the existence of variation in natural populations using gel electrophoresis, and later by using genetic sequencing to explore variation and the role of selection. Tooby and Cosmides are not oblivious to Lewontin's biological contributions to the study of natural variation. Here is what they say:

Modern geneticists, through innovative molecular genetic techniques, have certainly discovered within human and other species large reservoirs of genetic variability (Hubby and Lewontin 1966; Lewontin and Hubby 1966; see reviews in Ayala, 1976, and Nevo 1978). But it is only an adaptationist analysis that predicts and explains why the impact of this variability is so often limited in its scope to micro-level biochemical variation, instead of introducing substantial individuating design differences. (Tooby and Cosmides 1992, 79)

It is at least true that Lewontin and Hubby uncovered large reservoirs of genetic variation. The critical importance of their work was to demonstrate that there is substantial genetic variation in natural populations. That had been an issue for some time in early twentieth-century evolutionary biology. Hubby and Lewontin found a way to measure natural genetic variation. The technique they introduced, gel electrophoresis, was a method of revealing protein differences among organisms. The use of that method became a central feature in studies of natural variation before genetic sequencing became viable.

I said there were multiple ironies. Here is a second one. Evolutionary change depends on antecedent variation. That is beyond dispute. Tooby and Cosmides suggest, in the quote, above, that an "adaptationist analysis . . . predicts and explains" why the existing genetic variability does not issue in differences in fitness. What Tooby and Cosmides claim to "explain" is still a live issue. Lewontin explained why it is still at issue in a foundational book, *The Genetic*

Basis of Evolutionary Change (1974b). We certainly know that some genetic variation does not matter to adaptation. We also know that some does. In subsequent work, Lewontin and his students present empirical studies—in these cases, the genetic variation was assessed by genetic sequencing—that show the importance of variation, as well as empirical studies that show its unimportance, for specific differences in specific organisms. If empirical studies matter, they show that some genetic variability issues in differences in fitness, and some genetic variability does not issue in differences in fitness. Lewontin has been a major contributor to studies of natural populations, studies that are designed to determine when, or if, these differences matter for fitness. It turns out that they do sometimes matter. Whatever else is true, it is surely a mistake of the first order to claim that Lewontin is oblivious to the relevance of natural variation or to the importance of genetic variation to differences in fitness. Lewontin, to his credit, sees these as empirical issues. Tooby and Cosmides assume that the existing genetic differences do not matter. The empirical evidence is not at all consistent with that assumption. Some genetic variation does not matter, to be sure. Some does.

When we are told that human language and cognition are adaptations, we should ask what the alternative to this hypothesis might be. Pinker suggests at one point that the opposition to an adaptationist position is the view that human ingenuity is "a by-product of blood vessels in the skull that radiate heat as a runaway courtship device like the peacock's tail, as a stretching or chimpanzee childhood," among others (1987, 187). He is right that these and kindred explanations are badly "underpowered" (ibid.). He thinks that what is needed is the "leverage of good reverse engineering" (ibid.). If this is the opposition, Pinker could have written a much shorter book. I don't defend any of these as sufficient, though at least some could be articulated as alternatives. Buss (1999, 35–36) tells us that there are three theories concerning the origin of complex mechanisms: evolution by natural selection, creationism, and cosmic seeds (which he acknowledges is not really an alternative to selection). He sees that given such alternatives, there really is no contest. Natural selection is the only prospect for explaining human nature. Dennett's declared opponents in defending the view that our language is an adaptation are Gould and Chomsky, or in any case his rather extreme version of what he takes their views to be.[1] Dennett sides with Pinker and Bloom (1992), claiming that "although Gould has heralded Chomsky's theory of universal grammar as a bulwark against an adaptationist explanation of language, and Chomsky in return endorses Gould's antiadaptationism as an authoritative excuse for rejecting the obvious obligation to pursue an evolutionary explanation of the innate establishment of universal grammar, these two authorities are supporting each other over an abyss" (Dennett 1995, 391). It is an important and interesting question whether our linguistic capacities are adaptations. I expect they are. It is an equally important and interesting project to develop an evolutionary explanation of these capacities, whether or not they are adaptations. One might reserve skepticism over whether they are adaptations, and one might be skeptical about the specific adaptationist explanations proposed, without doubting for a moment that these are complex evolved traits. Even granting they are adaptations, there is a long gap between that and knowing what they are adaptations for.

The issue is not the importance of evolution or of natural selection, but the right explanation for evolutionary change. Pinker and Dennett argue as if the issue is whether there is evolution, or whether evolution can result from natural selection. Both claims are uncontroversial. They are also distinct. The conflation of the two is evident. Dennett's rendering of Darwin's "dangerous idea" blinds him to the difference. He says that Darwin's fundamental idea is that "Life on Earth has been generated over billions of years in a single branching tree—the Tree of Life—by one algorithmic process or another" (Dennett 1995, 51). Darwin, to the contrary, saw natural selection more specifically as the natural outcome of competitive struggle, driven by superfecundity. Evolution and speciation were the results, but Darwin's fundamental idea was the centrality of natural selection in the understanding of *adaptation*. Natural selection is one thing. Evolution is another. To treat Darwin's "fundamental idea" as Dennett does is to obscure the difference between evolution and evolution by natural selection.[2] Darwin carefully respected, and noted, the difference. The issue in understanding human evolution is mistakenly identified as whether there is evolution, or whether evolution *can* result from natural selection. Dennett and Pinker then feel confident in inferring that evolutionary psychology is secure: our psychological profile is properly explained as an adaptation to our Pleistocene heritage. The inference is an error. Neither evolution, nor the importance of natural selection is at issue. It is simply not enough to argue that a trait is evolved, and then conclude that it is an adaptation, much less that it is an adaptation to some particular environmental feature. For the argument that human psychology is the consequence of natural selection to be taken seriously, it cannot afford to gloss over the difference between evolution and its mechanisms. It cannot afford inattention to the details required of respectable evolutionary explanations. It cannot afford to ignore the importance of history, offering us evolution without history. What we require is a more developed and more articulated understanding of evolutionary biology if we are to shed light on the adaptive significance of human cognition and language.

3   The Prospects for an Evolutionary Psychology

I observed early on that evolutionary psychology offers a more modest agenda than did sociobiology, or at least apparently so. This said, the strength of the vision changes once the caveats are offered. Even after acknowledging the importance of "culture," Cosmides and Tooby say this at one point:

All humans share a universal, highly organized architecture that is richly endowed with contentful mechanisms, and these mechanisms are designed to respond to thousands of inputs from local situations. As a result, humans in groups can be expected to express, in response to local conditions, a variety of organized within-group similarities that are not caused by social learning or transmission. (1992, 116)

The conclusion they reach on these grounds is that "complex shared patterns that differ from group to group may be evoked by circumstances or may be produced by differential transmission" (ibid.) Instead of the initial focus on common behaviors across groups, we are left with a "universal, highly organized architecture." This is supposed to explain similarities within groups, and those similarities are, at least often, the consequence of our biology. If they are right that our cognitive architecture is uniform across the species, that will only obscure the case. Without evidence of variation, and the kind of variation, an occupation with evolution, much less selection, is pointless. The same dilemma faces Pinker and Bloom. It is because our linguistic capabilities are unique and universal that they find an evolutionary perspective compelling. However, if they are unique and universal, then we inevitably lack the kind of evidence that would allow us to construct a defensible evolutionary explanation.

The program advanced by advocates of evolutionary psychology is more modest in scope than that of sociobiology. It is nonetheless an ambitious one. The goal is not just to defend the idea that humans evolved, but to offer evolutionary explanations of specifically human features in terms of natural selection. These explanations are intended to reflect the evolutionary origins of the features they study. Human language and cognition are certainly complex traits. So is perception. So are our emotions and motivations. So are our hands. As with other traits, we may reasonably inquire into their evolutionary origins. The passions are also complex and human, though not specifically human, traits. Evolutionary psychologists offer us explanations of psychological mechanisms in terms of the specific environmental demands that shaped them in our evolutionary past. An understanding of these mechanisms and of their evolutionary functions should in turn illuminate the psychological mechanisms. I've emphasized the shortcomings of these evolutionary explanations, but there may be some hope for progress for some features of our psychological profile.

Here is a contemporary example that should feel familiar by this point. Feathers are an interesting feature of birds. They are exquisite adaptations. The fact that they are so elegantly fit to their aerodynamic functions is often appealed to in defending the idea that they are adaptations for flight. (See, e.g. Steiner 1917; Heilmann 1926; Parks 1966; Feduccia 1985, 1993, 1999.) The view is quite problematic. There are feathers, having many of the peculiarities of the familiar feathers on birds, present in fossils of theropod dinosaurs. Feathers evolved before the origin of birds, in these dinosaurs. What is worse, for the most part, these dinosaurs did not fly. The feathers on these ancestors have the typical branching structure of feathers. They have a tubular structure that is characteristic of feathers. They have the interlocking structures that give feathers the flat, but resistant, branched structure they need. (The flat structure with branches around a midline is called a "bipinnate" structure because feathers branch two ways around the midline. Each branch in turn has a set of interlocking barbs, which make feathers more rigid. This is a "pennaceous vane.") They are feathers. As Richard O. Prum and Alan H. Brush put it, "Concluding that feathers evolved for flight is like maintaining that digits evolved for playing the piano" (2002, 286). Indeed, as they show, the aerodynamic hypothesis gets the historical trajectory backwards. The feathers on theropod dinosaurs have a morphology that includes a bipinnate, closed pennaceous vane; moreover, this complex structure is a prerequisite for selection based on aerodynamic properties. That is, it is only after many of the structures are in place that selection could work to perfect them for flight.

The moral is clear enough. Feathers are not adaptations for flight, even though they do elegantly serve this purpose among birds. It would likely be wrong to expect a single selectionist trajectory behind the "evolution of feathers," as opposed to a number of different evolutionary events, one of which obviously found its way into the lives of birds. Reflection on the current function of feathers, and their role in flight, does not immediately shed light on the evolutionary history. A recognition of the relevant history is more important than speculations concerning their raison d'etre. The explanations wait on a description of the evolutionary history.

The turning point in our understanding not only of feathers, but of flight, was cladistic analyses. Instead of working with conjectured "primitive" forms or no forms at all, cladists generated analyses based on extant forms. Consistently, birds came up as descendants of small, carnivorous, theropod dinosaurs (see Padian 2001 and the references he provides). Kevin Padian says this:

Not long ago, decisions about phylogenetic relationships were often based on the behavioral, ecological, or functional plausibility that one taxon could have evolved from another. This judgment depended on the interpretation of structures and features in each taxon and suppositions about how evolutionary patterns would be expected to occur, given assumptions about how evolutionary processes should work. (2001, 599)

Cladistics changed the focus from direct ancestry to shared ancestry. Pattern became the key to descent. There are, of course, differences among avian biologists concerning the details of the evolution of birds. It is clear that feathers were characteristic of theropods and that birds are descendants within this group. It is also clear that birds (*Aves*) are suited for flight.

There is no comparable case to be made within evolutionary psychology in any general sense. Considered as an evolutionary hypothesis, the question is what the right explanation of human language or cognition is. The evolutionary psychologist's goal is to provide an evolutionary explanation of human traits as complex *adaptations*—as the products of natural selection—and not merely as *evolved* features of the human lineage. It is important to recognize that there is more than one possible evolutionary explanation even for complex characters, and not all would make the results adaptations. To know the evolutionary explanation requires knowing what something is an adaptation *for*—the conditions in response to which it evolved, among other things.

It is also important to recognize that even if natural selection plays a significant role, there will still be importantly different explanations possible. Once we recognize the contributions that evolutionary biology offers to evolutionary questions, the fact is that the resources of systematics and comparative biology offer little comfort to evolutionary psychology. Neither do the resources of population genetics. In both cases, the available biological and paleoanthropological record leaves the relevant ecological, social, phylogenetic, and populational variables sufficiently indeterminate that we do not know enough about the factors that shaped human evolution to evaluate the proposals offered by advocates of evolutionary psychology. We do know that language and intelligence are derived forms within the hominids, but it is not clear from the record whether they are features of *Homo sapiens*, or of a broader taxonomic class.[3] If we treat these as specifically human adaptations, for all we know we are focused on the wrong taxonomic level—as we did when fixing on feathers as adaptations for flight. We do know that social organization and the environment were relevant to the evolution of human language and intelligence, but we know little concerning the actual social structure or even the relevant physical environment. To be told that language facilitates communication is true, but that is hardly sufficient to develop an evolutionary explanation. To be told that human intelligence facilitates the management of information is also no doubt true, but that does not give any substance to an evolutionary explanation of our intellectual capacities.

On both sides of this discussion, rhetoric replaces reason. We should be cautious of the rhetoric that substitutes for more modest and reasonable claims others more ambitious and unsupported. The slippage is often striking. Here is an example, again from Philip Kitcher. In discussing the views of Richard Alexander, Kitcher turns to the case of primogeniture. It is common for elder children, and in particular for elder males, to receive a disproportionate amount of the wealth passed on to children. Alexander argued that this is to be explained in terms of the maximization of inclusive fitness. Kitcher notices that everything we know can be captured by the banal. Relatives do improve their station, and their fitness, by "mutual aid." Parents desire the welfare of their children, and siblings are often allies. The most obvious strategy for parents is to pass on their wealth to the eldest children. As Kitcher says, though, nothing in this guarantees that primogeniture maximizes inclusive fitness. The fact that we are allies with our kin and care for our children does not need that explanation. Moreover, that explanation garners no greater precision than the appeal to the banal. As Kitcher says, "the introduction of the evolutionary perspective brings with it no increase in predictive power. Folk psychology, unaugmented by any evolutionary ideas, will yield all the expectations Alexander claims for his own central hypothesis" (1985, 297).

Kitcher draws an analogous moral concerning the tendency of males, in some cultures, to support the children of his sisters rather than those of his wife. It is a system that might seem puzzling. Once again, Alexander explains it by appeal to inclusive fitness of the males involved. If there is substantial uncertainty concerning paternity, then it can in fact maximize inclusive fitness to preferentially support nephews and nieces rather than the offspring of one's wife. Even the vague and imprecise explanations of "folk psychology" get the same result. If one lives in a society in which there is such uncertainty—for example, one with substantial promiscuity—then the natural tendency to provide aid and comfort to one's children will not apply to the children of one's wife. So males turn their attentions to those with whom there is some certainty of kinship. Whatever vagueness may attach to such an explanation, it is enough. Nothing more in the way of precision is added by appealing to inclusive fitness.[4] This is, as Lewontin (1976) says, but "idle Darwinizing."

Notes

Introduction

1. Of course, natural selection might explain the reduction of these structures through what Darwin called "economy of growth." This was wholly clear to Darwin. Explaining the reduction of the structures (why male nipples are small) is not the same as explaining why the structures are retained in a reduced form (why they are present at all).

2. Darwin and Wallace parted company at just this point, with Wallace pressing for an expanded role for natural selection. For useful discussions of the historical issues, see Richards 1987, ch. 6; Cronin 1991, ch. 2; Kottler 1985; and Browne 2003, ch. 9.

3. Again, my discussion will be lamentably brief. For insightful discussions, see Richards 1987, chap. 6; also Desmond and Moore 1991, 264ff.

4. There are numerous decisive criticisms of creationist views. Some particularly worth recommending include Niles Eldredge's *The Triumph of Evolution and the Failure of Creationism* (2000), Douglas Futuyma's *Science on Trial: The Case for Evolution* (1983), and Philip Kitcher's *Abusing Science: The Case Against Creationism* (1982).

5. There are interesting and important issues here. Finding a probabilistic connection between two variables goes two ways. Which is probative as evidence for a causal connection is critical. A useful example is provided by evidence concerning XYY chromosomes and the tendency to violence. Richard Speck had an XYY chromotype and was a multiple murderer. That was used in his defense since many multiple murderers had an XYY chromotype. The question is whether the XYY chromotype is predictive (or causative) of homicidal tendencies. The evidence used concerned whether serial murderers tend to have an XYY chromotype. That evidence is not directly relevant to the first question. What we need to know is whether being XYY is predictive of homicide, and not whether killers are XYY. (The other direction, predicting homicide from XYY chromotypes involves very small numbers and, in any case, is not the direction we want to know about.) It was commonly concluded that XYY chromotypes tended to be multiple murderers, even though it turns out that most XYY chromotypes are just normal people.

Chapter 1

1. I am indebted to George Uetz for this recounting of the actual frequency of envenomation in spiders and the risks they pose. Bruce Jayne similarly enlightened me concerning snakes and their associated risks.

2. Buss is right that evolutionary scientists hold that adaptations are the products of natural selection, though that is essentially a terminological point. A trait, however complex, that was not selected for would not be thought of as an adaptation. The interesting thought is certainly that complex features are adaptations because they had to be the products of natural selection.

3. Buller (2005), discusses the issue in chap. 4 of *Adapted Minds*. In this chapter which derived from other work with Valerie Hardcastle (2000), he presses that the brain is better understood as a general-purpose problem solver.

4. The point that we cannot readily articulate social assumptions characteristic of our culture is also irrelevant to the issue. It is equally difficult for us to articulate the principles involved in speaking English. Native speakers of a language can deploy such principles readily, and though they are not readily articulated (without the aid of linguists), some are certainly learned.

5. At the London School of Economics (LSE), John Ashworth established the Centre for the Philosophy of the Social and Natural Sciences as a conduit for evolutionary psychology in the 1990s. Helena Cronin was an early appointment, and the Centre sponsored sociobiologists such as Richard Dawkins, Matt Ridley, and Robert Trivers, as well as evolutionary psychologists such as Margin Daly, Steven Pinker, and Margo Wilson. The connections are institutional and not accidental. On the other side, some of the most prominent sociobiologists and their opponents are residents at the Museum of Comparative Zoology at Harvard.

6. The most commonly discussed case is undoubtedly children raised in Israeli kibbutzim (see van den Berghe 1983).

7. The point was made by Lewontin, Rose, and Kamin (1984). That it is obvious does not make it less important. Lumsden and Wilson (1981) attempt to answer the point, but unsuccessfully (cf. Kitcher 1985, 347–394). Humans are generally averse to consuming putrid foods. We are not averse, or at least many are not averse, to eating pork. The former may be due to biological features, though it is important to ask, for example, why dogs do not share our aversions, even though they are not all that distant biologically. We could have as readily evolved the aversion to putrid food as we could have evolved a tolerance for it. The aversion to pork needs a decidedly different form of explanation.

8. In the case of the !Kung, apparently, bands are also not organized around biological patterns, and marriage between close biological relatives is not at all uncommon. It was not even uncommon in Victorian England.

9. It is, of course, the very task that hereditarian analyses undertake, though they all know phenotype is a function of both genes and environment.

10. It is wildly irrational to avoid potentially therapeutic treatments when diseases are fatal. This was the point of protests against drug therapies for HIV, an otherwise fatal infection.

11. There are negative costs to treatments since the number of treatments is limited. If one were to receive a heart transplant, that would be only at the expense of others, given the limit in the number of hearts. Costs aside, these are difficult issues.

12. The distinction was originally due to F. Knight (1921). In the technical literature, a choice involves *risk* provided the indeterminacies attending choice can be captured in terms of specific probabilities. There is risk attached to games of chance, such as roulette. By contrast, a choice involves *uncertainty*, provided the indeterminacies from a choice are not assigned objective probabilities. There have been dramatic advances in the understanding of choice under uncertainty, but decision under risk has the advantage of being able to draw on the resources of the theory of probability.

13. Philip Kitcher and A. Leah Vickers (2002) offer a devastating critique of the reasoning behind Thornhill and Palmer's view. They point out that a Darwinian account of rape would be expected to issue an explanation based on at least an assessment of the relative fitness of the behavioral strategies (and that would require a relatively explicit statement of the strategies, which Thornhill and Palmer do not offer). As Kitcher and Vickers point out, Thornhill and Palmer offer no guidance to assess the "benefits" to a male inclined toward coercive sex, or to the "costs." Copulation infrequently leads to pregnancy, and the progeny of rape might readily be abandoned—as might be the female. In ancestral groups, the "costs" could be quite high, given the reactions of others in the social group. Ostracism would be mild as a punishment, though that would often have been fatal. Instead of serious analysis, Kitcher and Vickers conclude we're offered only "banal" interpretations (see Kitcher 2003, 244 ff.).

Chapter 2

1. Generally, the architects of evolutionary psychology resist such an ahistorical approach to the study of adaptation, but the method has left its mark and certainly has a salient place in biology, including the work of R. A. Fisher (1930) in population genetics and that of R. H. MacArthur (1957, 1960) in evolutionary ecology. It has been particularly influential in studies of animal behavior (e.g., see Alcock 2001).

2. I favor the view held by Ron Amundson, Peter Godfrey-Smith, and George Lauder that there are at least two legitimate uses of "function" within biology. The first is the historical view that "proper" functions are what was selected for in the past—the products of natural selection in the sense that they previously were selected *for*. The other is more common among anatomists and physiologists than with evolutionary biologists and looks to current causal role as definitive of function. The issues I raise here survive either reading, though the historical use would make things more difficult for reverse engineering. It should be clear from the context which account of function is intended, though when there is ambiguity, I will explicitly differentiate between *historical* and *physiological* function. For one interested in pursuing the various issues this raises, I recommend Paul Davies's *Norms of Nature* (2001).

3. Dennett has been brilliant in uncovering philosophical pretense. One theme I applaud is the defusing of thought experiments. For a wonderful treatment of the pretense, see his "Quining Qualia" (1988). He systematically urges philosophers toward reality and away from fantasy. I have learned from Dennett and embrace the vision. Our differences are over the real rather than the imaginary.

4. For other applications, see Pinker and Bloom 1992 and Shepard 1992.

5. The London specimen was discovered by H. von Meyer in 1861. Richard Owen acquired the specimen from Solnhofen for the British Museum of Natural History in 1862 (see Owen 1862). For whatever reason, Huxley did not examine it for another five years. *Archaeopteryx*, he allowed, was a bird, but with reptilian features. Huxley's views were set forth in "On the Animals," before the Royal Society (Huxley 1868). For useful discussions, see Desmond 1984, 1994 and Lyons 1999.

6. Archosauria is a broad group, including not only dinosaurs and birds, but crocodilians. Thecodonts were agile and quick, with legs beneath their bodies. Dinosaurs evolved from thecodonts during the late Triassic.

7. Cuvier (1812) thought there were four fundamental *Baupläne*, characterizing the major animal phyla or *embranchements*: Vertebrates (e.g., humans), Mollusks (e.g., an octopus), Articulates (e.g., insects), and Radiata (e.g., starfish). We now recognize more than thirty animal phyla.

8. Gould and Lewontin have been broadly criticized and sometimes disparaged. Evolutionary psychologists are certainly not among their friends. It would be tedious to survey the critical literature, and I won't do so. For one who wants to explore the literature, some of the critics include: Alcock 1998; Alcock 2001; Borgia 1994; Buss et al. 1998; Cronin 1991; Dawkins 1986; Dennett 1995, 1997; Maynard Smith 1978, 1995; Mayr 1983; Pinker 1997; Pinker and Bloom 1992; Reeve and Sherman 1993; Sherman 1998 and 1999; Thornhill 1990; and Tooby and Cosmides 1992. The critics by no means share the same objections, and for the resolute it at least makes amusing reading.

9. It is interesting that Buss et al. (1998) spend a great deal of time complaining about the "confusions" in Gould's view, though the key point Gould is pressing—that current function is not the same as evolutionary function—seems to be a view they endorse.

10. As a somewhat technical matter, this should require, not merely that there be differences between males and females on average, but that the variance within groups should not swamp the average difference between groups. Given the focus on differences between average values in this literature, and lacking any evident concern with variance within groups, speaking of the difference between males and females indicates only a difference between average values. The within group variation among males and the variation among females are no less relevant. I've seen no evidence of this sort presented. Perhaps it is available and perhaps not. Buller does seem to notice

this issue; partly in response, he offers some alternatives to the simple mate choice models offered in the literature. His alternatives are no less speculative than Buss's, and Buller concedes there is insufficient evidence to support either view. I agree.

11. It is difficult to use this approach generally. Such a priori optimization requires substantive knowledge of the environmental "problem" to be solved, the range of phenotypic variation, their relative efficiency, and much more (cf. Lauder 1996). It is not impossible to apply it, as we'll see.

12. This would presumably mean heterogeneity of the selective environment, in Brandon's scheme. Notice this could be seen even without knowing anything of the relevant physical environment.

13. As Gould and Lewontin observed, this kind of retrofitting does ensure that there will be some optimal model for the behavior in question. Kingsland (1985) illustrates this moral with some stunning examples involving population growth. Responses to the challenge of finding an analysis that makes a trait optimal depend primarily on the creativity of the theorist.

14. I take my attitude to be wholly consonant in this respect with those of Gould and Lewontin, both distinguished evolutionary biologists. Here we actually do not differ from advocates of evolutionary psychology, including Ketelaar and Ellis.

15. The quotation within Ospovat is to Darwin's B notebook, p. 84.

16. Notice this is exactly what Darwin appeals to in both the *Origin* and in the *Descent*. The insight was one that Darwin carried throughout his career.

17. The requirement is evident if one consults the *Sketch* of 1842 or the *Essay* of 1844, though it is less clear in the argument of the *Origin* (1859). See Hodge 1977 and Ruse 1979b.

18. It is worth noting that Hume was interested in pointing out a limitation on such reasoning. The ideal of a *vera causa* is more ambitious, aimed at defining the conditions under which an explanation—or an appeal to a cause—is adequate. Humeans should feel reasonably comfortable with the more ambitious limits.

Chapter 3

1. We could count these as "rules" if we want. So there is a rule to avoid spiders and snakes if evolutionary psychologists are right. These are treated very nearly as "reflexes." There is certainly a reflex to blink when an object moves quickly toward the eye. That too embodies a "rule." Strictly, reflexes are mediated by connections in the spinal column, so only the latter is properly a reflex.

2. He also incorporates some discussion of the social functions of rationality. This leads to the suggestion "that rationality is shaped, selected, and maintained not to serve a level below that of organisms but to serve a level *above* the level of institutions" (Nozick 1993, 126). I do not pretend to know what this might mean.

3. Davies, Fetzer, and Foster (1995) effectively defuse the arguments for domain specificity offered by Cosmides and Tooby. They point out that Cosmides and Tooby falsely assume, for example, that the general rules would be couched in terms of first-order logic, whereas there are many effective and plausible alternatives for general formal rules.

4. See, among those for modularity, Carruthers 1992, 2003; Cosmides and Tooby 1992; and Gigerenzer and Hug 1992. See, among those against modularity, Dupré 2001; Currie and Sterelny 1999; and Buller 2005.

5. Even this way of putting the issue gives too much to Cosmides and Tooby. The very idea of "how much variance" in any absolute sense itself makes little sense.

6. Within population genetics, the response to selection is simply the product of the intensity of selection and the heritability, in the narrow sense that includes only additive genetic variance.

7. This simplifies matters somewhat, in ways that do not matter just yet. I'll add the needed qualifications in chap. 4, where phylogenies occupy center stage.

8. This has sometimes been ingeniously addressed, by using experimental settings.

9. There is another allele relevant to sickle cell resistance, though it is largely unrepresented in the populations and often unmentioned in discussions of balancing selection. The C allele is a recessive form that confers enhanced resistance without the deleterious effects of the S allele. Because it is expressed only in a homozygous form, the advantages it offers are rarely expressed; if the population begins with C at low frequencies, it does not increase. Nonetheless, the fitness maximum for a population with the three alleles available occurs with $f(C)$ at 1.00. In this larger context, it would be wrong to conclude that natural selection is a "sufficient cause." It would be equally wrong to conclude that natural selection played no role: natural selection is *necessary* for adaptive evolution, but it is *not sufficient* to define an adaptive process (see Templeton 1982).

10. Similarly, *Astyanax fasciatus*, the Mexican cave fish, underwent regressive evolution more than once, when different surface populations were isolated (see Mitchell, Russell, and Elliott 1977).

11. Humans have some relatively unique features, including a larynx that is lower than that even of fossil hominids. This would in turn affect the range of distinguishable vocalizations, especially vowels, which depend on an enlarged pharynx.

12. I will review some of the evidence for these paleoanthropological claims in the next chapter. This is a field that changes dramatically and quickly.

13. More recent work has met with considerable results, from Washoe to Kanzie. I think these are testimonies to the intelligence of our primate relatives, but see no reason to think that they offer special insights into human language learning or use.

14. Songbirds have a left hemisphere specialization for song production, and Japanese macaques have a significant left hemisphere involvement in recognition of calls. It is not clear whether asymmetries for language are primitive or derived.

15. Here I oversimplify considerably. Much depends on the specific sorts of learning rules that are allowed; even more depends on the sorts of constraints on learning that are imposed. Deacon (1997) discusses many of the complexities raised by these sorts of considerations. In the end, he allows that the constraints on language evolution lie in the brain and in development. He contends that this tells against Chomsky, but doesn't finally come to grips with the question whether there are sufficient cognitive constraints to allow for language learning in the absence of specifically linguistic constraints on acquisition.

16. I sincerely doubt that the genetic variance will actually turn out to be zero. Behavioral geneticists repeatedly uncover significant heritability values; and evolutionary biologists find genetic variation almost anywhere they look. I suppose that evolutionary psychologists suggest zero heritability because they assume there has been persistent directional selection.

17. It also turns out that there is an enhanced planum termporale among chimpanzees (Gannon et al. 1998). This is the language area in the human brain, a component of Wernicke's area. This complicates the issues in multiple ways, especially in light of the fact that chimpanzees do not have sophisticated linguistic capabilities.

18. We should be careful not to assume that this is a lineage. It is, instead, a group of related species, some of which clearly overlapped in time.

19. There is a considerable industry concerned with the measurement of intelligence (IQ), and with assessing its heritability. I think there are many reasons to be skeptical about the enterprise, but it is in any case tangential to the concerns here. Cosmides and Tooby emphatically reject any approach to cognition that embraces generalized learning capacities, as is common in the IQ industry.

20. I suspect that this is the source of Chomsky's skepticism. In fact, though I have no doubt about evolution or the importance of natural selection, even in the case of linguistic abilities, I take the morals I am pressing to be concordant with Chomsky's views.

21. Cosmides and Tooby subsequently appeal to contemporary nonagricultural groups in support of their conclusions. Such groups do not represent the ancestral hominids, but are alternative human cultural variants. The fact that such societies follow the general conclusions from the

formal models is of no significance. The evolutionary models require that there be genetic variance for such differences in ancestral populations. There is no evidence whatsoever to think that the Ache and San differ for any genetic factors relevant to their differences; more plausibly, these are cultural adaptations to different, but extreme, conditions. There is no evidence whatsoever for the kind of genetic variation that Cosmides and Tooby require.

Chapter 4

1. It is not clear to me how comfortably this conception of the function of reason fits with the modular architecture so common among evolutionary psychologists. The roots of the modular conception lie with the "organology" of Gall, and are opposed, at least historically, to the more Cartesian picture of the function of rationality, a picture explicitly repudiated by Cosmides and Tooby as the "standard social science model." This would be an interesting thread to follow, but not one I will follow here. See Looren de Jong and van der Steen 1998, sec. 4, for some discussion of this view.

2. It is often thought, mistakenly, that this would preclude traits maintained by stabilizing selection. That it does not is easily seen. If a trait is present due to stabilizing selection, then it must have been maintained in a lineage by selection. That too is a comparative issue, depending on whether, in phylogenetic ancestors, there was variation or whether, in phylogenetic ancestors, that variation would have been maintained. It is also directly relevant to determine whether environments have been stable.

3. Such phylogenetic trees should be based on a large number of characters, and when they are, they are very robust. Including the characters at the focus of an analysis is generally harmless since the topology of a tree will not depend critically on any one character.

4. Cladism is a particular approach to constructing and using phylogenetic trees. It is controversial. In general, I'll describe a cladistic methodology, though alternatives would suffice to make essentially the same point.

5. The same points could be made using continuous variables, though the analysis is more complicated.

6. Choice of an outgroup is difficult and important. It cannot be among the species being considered, but needs to be closely related if it is to offer useful information. If it is possible to use a close relative, then that provides the tool for deciding which characters were present in the ancestors to the clade and which provide useful phylogenetic information. As I've said, statistical testing faces the same problem. We need an appropriate contrast class, which will inform us about the structure of the trees; and a wrong choice can substantially mislead us.

7. It is, of course, "derived" insofar as it is inherited. A *derived* trait in this scheme is one that changed within the lineage. The chin is a derived trait for *Homo sapiens*. An *ancestral* trait is one that is inherited, but present in the lineage. The pentadactyl structure of our hand is ancestral.

8. Once we accept that some characters may evolve more than once, there are a number of procedures for selecting the best tree (see Felsenstein 1982 for an extensive discussion). If one allows derived characters to evolve more than once, the most parsimonious tree might be the one with the fewest character state derivations. If we hold that complex characters cannot evolve more than once but can be lost repeatedly, the most parsimonious tree would be that with the fewest losses of derived characters. We might hold simply that the most parsimonious tree is the one with the fewest character state transitions. These are important issues for systematists. The various criteria result in significantly different descent frees, but such disputes do not centrally affect the points I will be making here.

9. A trait may be ancestral and still require an evolutionary explanation if we know, for example, that variations arose regularly. Since we would expect that in the absence of selection, common variations would sometimes be established in a set of lineages, if we know independently that there is variation for a trait, then we might properly explain the ubiquity of a trait in terms of selection. A phylogenetic analysis can also reveal information about the extent of variation.

10. The topology of the tree for the family Rhinoceratidae is similar to our toy example. If species D represents the one-horned form and the remaining species are all two-horned, then what we can see is that the one-horned form is a derived form; what we would then seek is an explanation of the one-horned variety as evolved from an ancestral two-horned form.

11. This does not mean that humans evolved from any of the modern apes. Humans and apes evolved from a common ancestor. There are some fossil remains from the Miocene that are plausible candidates for a common ancestor, and it is the record in the late Miocene and early Pliocene that is especially lacking. This is evidently the critical period for the divergence between apes and hominids.

12. This is a bit of an oversimplification. As with nearly everything else in human evolution, there is disagreement. Gait needs to be inferred from functional interpretations—considerations of "design." Sometimes we do not have the postcranial materials we would like, and even when we do, there are commonly alternative interpretations. Some hominids were likely adept climbers, for example. On the other hand, some lived in woodlands, which tells against the view that bipedalism evolved in the savanna.

13. These robust australopithecines share a number of cranial features, probably related to diet. In part, this leads some paleoanthropologists to assign them to a distinct genus, *Paranthropus*. However, these craniodental features are evidently not independent developmentally, and so cladistic analyses that treat them as independent may overweight them. For interesting discussions from a cladistic perspective, see Skelton and McHenry 1992, 1997; Lieberman 1995; Lieberman, Wood, and Pilbeam 1996; and McCollum 1999.

14. This is a confusing literature. Many cladograms have been offered. Some imply that the genus *Homo* is paraphyletic, and thus not a legitimate group.

15. There has been striking work on the evolution of brain size (e.g., Jerison 1973), though it is not without problems (e.g., Gould 1996). Even if endocranial capacity were a reliable measure of intelligence, the measures of endocranial capacity are problematic. More recent computerized scans of an *A. africanus* skull were used to generate a three-dimensional computer model. The resulting value for endocranial volume was substantially lower than previous more conventional estimates (518 cc as opposed to values greater than 600 cc). This opens the prospect that endocranial values have been systematically overestimated. See Conroy et al. 1998 for discussion.

16. If Deacon is right that human intelligence and human language coevolved, this is exactly what we should expect. This raises a number of interesting issues, discussed at length by Deacon (1997) in *The Symbolic Species: The Co-evolution of Language and the Brain*.

17. By contrast, there is no support for the thought that the human brain includes structures wholly lacking in other primates, though there undoubtedly are modifications in architecture specific to humans.

18. In other cases, as when we turn to parental care, family structure, or the moral sentiments, we can be certain that a concentration on *Homo sapiens* is much too narrow; in one form or another, these characterize the hominids more generally and often our primate cousins.

Chapter 5

1. His interpretation is a distortion of both Chomsky and Gould. So far as I can tell from either of their writings, or from the passages Dennett cites, their position is more moderate and more defensible than Dennett recognizes. Of course, the ideal opponent, rhetorically, is one whose position is extreme and untenable. Strategic advantage is no defense for rhetorical excess.

2. Dennett's "algorithms" have guaranteed results, executed consistently. This means that evolution (or perhaps natural selection) would have deterministic consequences. Evolutionary explanations, though, are fundamentally probabilistic. I have not emphasized the probabilistic character of evolutionary theory here, but it is absolutely central in understanding any of the discussions concerning the significance of natural selection. It would be possible to introduce probabilistic algorithms, of course, but that would require that Dennett abandon some of his criticisms of Gould's commitment to contingency (cf. Dennett 1995, chap. 10).

3. See note 18 to chaper 4, above.

4. Kitcher goes even further to show that it is not at all clear that an appeal to inclusive fitness yields the expected result. In point of fact, whether this would be expected depends on much that is not specified and not known. Under some specifications of the case, an avunculate system would be unstable from an evolutionary perspective. Whether a specification depending on inclusive fitness is even consistent with this system depends on assumptions about the range of alternative strategies, the specific social practices that prevail, and much more. The ethnographic record is just not good enough to support such specificity. See Kitcher 1985, 299–307.

References

Abrams, P. 2001. "Adaptationism, Optimality Models, and Tests of Adaptive Scenarios." In Orzack and Sober 2001, 273–302.

Adler, J. E. 1984. "Abstraction Is Uncooperative." *Journal for the Theory of Social Behaviour* 14:165–181.

Alberch, P. 1982. "Developmental Constraints in Evolutionary Processes." In J. T. Bonner, ed., *Evolution and Development, Dahlem Conference Report No. 20*, 313–332. New York: Springer Verlag.

Alcock, J. 1988. "Unpunctuated Equilibrium in the Natural History Essays of Stephen Jay Gould." *Evolution and Human Behavior* 19:321–336.

Alcock, J. 2001. *The Triumph of Sociobiology*. Oxford: Oxford University Press.

Alexander, R. D. 1979. *Darwinism and Human Affairs*. Seattle: University of Washington Press.

Alexander, R. D. 1989. "Evolution of the Human Psyche." In P. Mellars and C. Stringer, eds., *The Human Revolution*. Edinburgh: University of Edinburgh Press.

Alexander, R. D. 1990. "How Did Humans Evolve?" Special Publication 1, Museum of Zoology. Ann Arbor: University of Michigan Press.

Allen, E., and the Sociobiology Study Group. 1976. "Sociobiology—Another Biological Determinism." *BioScience* 26:182, 184–186.

Amundson, R. 1994. "Two Concepts of Constraint: Adaptation and the Challenge from Developmental Biology." *Philosophy of Science* 61:556–578.

Amundson, R. 1996. "Historical Development of the Concept of Adaptation." In Rose and Lauder 1996, 11–54.

Amundson, R. 1998. "Typology Reconsidered: Two Doctrines in the History of Evolutionary Biology." *Biology and Philosophy* 13:153–177.

Amundson, R. 2001. "Adaptation and Development: On the Lack of Common Ground." In Orzack and Sober 2001, 303–334.

Amundson, R., and G. V. Lauder. 1994. "Function without Purpose: The Uses of Causal Role Function in Evolutionary Biology." *Biology and Philosophy* 9:443–469.

Anderson, J. R. 1991. "Is Human Cognition Adaptive?" *Behavioral and Brain Sciences* 14:471–517.

Andrews, P., and L. Martin. 1987. "Cladistic Relationships of Extant and Fossil Hominoids." *Journal of Human Evolution* 16:101–118.

Antonovics, J., and A. D. Bradshaw. 1970. "Evolution in Closely Adjacent Plant Populations. VII. Clinal Patterns at a Mine Boundary." *Heredity* 25:349–362.

Antonovics, J., A. D. Bradshaw, and R. G. Turner. 1971. "Heavy Metal Tolerance in Plants." *Advances in Ecological Research* 7:1–85.

Ardrey, R. 1966. *The Territorial Imperative*. New York: Atheneum.

Arnold, S. J. 1992. "Constraints on Phenotypic Evolution." *American Naturalist* 140 (suppl.): S85–S170.

Arnold, S. J. 1994. "Multivariate Inheritance and Evolution: A Review of the Concepts." In C. R. B. Boake, ed., *Quantitative Genetic Studies of Behavioral Evolution*. Chicago: University of Chicago Press, 17–48.

Asfaw, B., T. White, O. Lovejoy, B. Latimer, S. Simpson, and G. Suwa. 1999. "*Australopithecus garhi*: A New Species of Early Hominid from Ethiopia." *Science* 284:629–635.

Axelrod, R. 1984. *The Evolution of Cooperation*. New York: Basic Books.

Axelrod, R., and W. D. Hamilton. 1981. "The Evolution of Cooperation." *Science* 211:1390–1396.

Barash, D. 1976. "Male Response to Apparent Female Adultery in the Mountain Bluebird: An Evolutionary Interpretation." *American Naturalist* 110:1097–1101.

Barash, D. 1977. *Sociobiology and Human Behavior*. New York: Elsevier.

Barash, D. 1979. *The Whisperings Within*. London: Penguin.

Barkow, J. H., L. Cosmides, and J. Tooby, eds. 1992. *The Adapted Mind: Evolutionary Psychology and the Generation of Culture*. New York and Oxford: Oxford University Press.

Baron, J. 1994. *Thinking and Deciding*, 2nd ed. Cambridge: Cambridge University Press.

Baron-Cohen, S. 1995. *Mindblindness: An Essay on Autism and Theory of Mind*. Cambridge, Mass.: MIT Press.

Barr, T. C. 1968. "Cave Ecology and the Evolution of Troglobites." *Evolutionary Biology* 2:35–102.

Baum, D. A., and A. Larson. 1991. "Adaptation Reviewed: A Phylogenetic Methodology for Studying Character Macroevolution." *Systematic Zoology* 40:1–18.

Beatty, J. 1980. "Optimal-Design Models and the Strategy of Model Building in Evolutionary Biology." *Philosophy of Science* 47:532–561.

Bechtel, W., and R. C. Richardson. 1993. *Discovering Complexity: Decomposition and Localization as Strategies in Scientific Research*. Princeton: Princeton University Press.

Benton, T. 2000. "Social Causes and Natural Relations." In Rose and Rose 2000, 249–271.

Berkeley, D., and P. Humphreys. 1982. "Structuring Decision Problems and the 'Bias Heuristic.' " *Acta Psychologica* 50:201–252.

Boake, C. R. B., S. J. Arnold, F. Breden, L. M. Meffert, M. G. Ritchie, B. J. Taylor, J. B. Wolf, and A. J. Moore. 2002. "Genetic Tools for Studying Behavior." *American Naturalist* 160d:S143–S159.

Bock, W. J. 1965. "The Role of Adaptive Mechanisms in the Origin of the Higher Levels of Organization." *Systematic Zoology* 14:272–287.

Bock, W. J. 1986. "The Arboreal Origin of Avian Flight." *Memoirs of the California Academy of Science* 8:57–72.

Bodmer, W. F., and L. L. Cavelli-Sforza. 1976. *Genetics, Evolution, and Man*. San Francisco: W. H. Freeman.

Borgia, G. 1994. "The Scandals of San Marco." *Quarterly Review of Biology* 69:373–375.

Brandon, R. 1978. "Adaptation and Evolutionary Theory." *Studies in History and Philosophy of Science* 9:181–206.

Brandon, R. 1990. *Adaptation and Environment*. Princeton: Princeton University Press.

Brandon, R., and M. D. Rausher. 1996. "Testing Adaptationism: A Comment on Orzack and Sober." *American Naturalist* 148:189–201.

Brooks, D. R., and D. A. McLennan. 1991. *Phylogeny, Ecology, and Behavior: A Research Program in Comparative Biology*. Chicago: University of Chicago Press.

Browne, J. 2002. *Charles Darwin: The Power of Place*. New York: Alfred A. Knopf.

Buller, D. J. 1999. "DeFreuding Evolutionary Psychology: Adaptation and Human Motivation." In Hardcastle 1999, 99–114.

Buller, D. J. 2005. *Adapting Minds: Evolutionary Psychology and the Persistent Quest for Human Nature.* Cambridge, Mass.: MIT Press.

Buller, D. J., and V. Hardcastle. 2000. "Evolutionary Psychology Meets Developmental Neurobiology: Against Promiscuous Modularity." *Mind and Brain* 1:307–325.

Burian, R. 1983. "Adaptation." In M. Grene, ed., *Dimensions of Darwinism.* Cambridge: Cambridge University Press, 287–314.

Burian, R. 2004. *Epistemological Essays on Development, Genetics, and Evolution.* New York: Cambridge University Press.

Burian, R., and R. C. Richardson. 1991. "Form and Order in Evolutionary Biology: Stuart Kauffman's Transformation of Theoretical Biology." In M. Forbes and A. Fine, eds., *PSA 1990.* Vol. 2. East Lansing, Mich.: Philosophy of Science Association, 267–287.

Burian, R., and R. C. Richardson. 1992. "A Defense of Propensity Interpretations of Fitness." In M. Forbes and D. Hull, eds., *PSA 1992.* Vol. 1. East Lansing, Mich.: Philosophy of Science Association, 1992, 349–362.

Burkhardt, F., ed. 1983–2005. *The Correspondence of Charles Darwin.* 15 vols. 1821–1864. Cambridge: Cambridge University Press.

Buss, D. M. 1994. *The Evolution of Desire: Strategies of Human Mating.* New York: Basic Books.

Buss, D. M. 1995. "Evolutionary Psychology: A New Paradigm for Psychological Science." *Psychological Inquiry* 6:1–30.

Buss, D. M. 1999. *Evolutionary Psychology: The New Science of Mind.* Boston: Allyn and Bacon.

Buss, D. M. 2000. *The Dangerous Passion: Why Jealousy Is as Necessary as Love and Sex.* New York: The Free Press.

Buss, D. M., M. Haselton, T. K. Shackelford, A. L. Bleske, and J. C. Wakefield. 1998. "Adaptations, Exaptations, and Spandrels." *American Psychologist* 53:533–548.

Buss, D. M., R. J. Larsen, and D. Westen. 1996. "Sex Differences in Jealousy: Not Gone, Not Forgotten, and Not Easily Explained by Alternative Hypotheses." *Psychological Science* 7:373–375.

Buss, D. M., R. J. Larsen, D. Westen, and J. Semmelroth. 1992. "Sex Differences in Jealousy: Evolution, Physiology, and Psychology." *Psychological Science* 3:251–255.

Buss, D. M., and T. K. Shackelford. 1997. "From Vigilance to Violence: Mate Retention Tactics in Married Couples." *Journal of Personality and Social Psychology* 72:346–351.

Buss, D. M., T. K. Shackelford, L. A. Kirkpatrick, J. Choe, M. Hasegawa, T. Hasegawa, and K. Bennett. 1999. "Jealousy and the Nature of Beliefs about Infidelity: Tests of Competing Hypotheses About Sex Differences in the United States, Korea, and Japan." *Personal Relationships* 6:125–150.

Caplan, A. L., ed. 1978. *The Sociobiology Debate: Readings on the Ethical and Scientific Issues Concerning Sociobiology.* New York: Harper and Row.

Carruthers, P. 1992. *Human Knowledge and Human Nature.* Oxford: Oxford University Press.

Carruthers, P. 2003. "On Fodor's Problem." *Mind and Language* 18:502–523.

Chagnon, N. A. 1968. *Yanomamö: The Fierce People.* New York: Holt, Rinehart, and Winston.

Chagnon, N. 1983. *Yanomamö: The Fierce People,* 3rd edition. New York: Holt, Rinehart, and Winston.

Chagnon, N. A. 1988. "Male Yanomamo Manipulations of Kinship Classifications of Female Kin for Reproductive Advantage." In L. Betzig, M. Borgerhoff Mulder, and P. Turke, eds., *Human Reproductive Behavior: A Darwinian Perspective,* 23–48. New York: Cambridge University Press.

Chamberlain, A. T., and B. A. Wood. 1987. "Early Hominid Phylogeny." *Journal of Human Evolution* 16:119–133.

Chambers, R. 1847. *The Vestiges of the Natural History of Creation*, 6th edition. London: Churchill. Originally published anonymously.

Chen, P., Z. Dong, and S. Zhen. 1998. "An Exceptionally Well-Preserved Dinosaur from from the Yixian Formation of China." *Nature* 391:147–152.

Chomsky, N. 1972. *Language and Mind.* Enlarged ed. New York: Harcourt, Brace Jovanovich.

Chomsky, N. 1980. *Rules and Representations*. New York: Columbia University Press.

Christiansen, K. A. 1965. "Behaviour and Form in the Evolution of Cave Collembola." *Evolution* 19:529–532.

Coddington, J. A. 1988. "Cladistic Tests of Adaptationist Hypotheses." *Cladistics* 4:5–22.

Cohen, L. J. 1986. *The Dialogue of Reason*. Oxford: Oxford University Press.

Conroy, G. C., G. H. Weber, H. Seidler, P. V. Tobins, A. Kane, and B. Brunsden. 1998. "Endocranial Capacity in an Early Hominid Cranium from Sterfontein, South Africa." *Science* 280:1730–1731.

Cosmides, L., and J. Tooby. 1987. "From Evolution to Behavior: Evolutionary Psychology as the Missing Link." In J. Dupré, ed., *The Latest on the Best: Essays on Evolution and Optimality*, 277–306. Cambridge, Mass.: MIT Press.

Cosmides, L., and J. Tooby. 1992. "Cognitive Adaptations for Social Exchange." In Barkow, Cosmides, and Tooby 1992, 163–228.

Cosmides, L., and J. Tooby. 1994. "Origins of Domain Specificity: The Evolution of Functional Organization." In L. A. Hirschfeld and S. A. Gelman, eds., *Mapping the Mind: Domain Specificity in Cognition and Culture*. Cambridge: Cambridge University Press. Reprinted in R. Cummins and D. D. Cummins, eds., *Minds, Brains, and Computers*. Oxford: Blackwell, 2000, 523–543.

Cosmides, L., and J. Tooby. 1997. "The Modular Nature of Human Intelligence." In A. B. Scheibel and J. W. Schopf, eds., *The Origin and Evolution of Intelligence*. Sudbury, Mass.: Jones and Bartlett, 71–101.

Cronin, H. 1991. *The Ant and the Peacock*. Cambridge: Cambridge University Press.

Cronin, H. 2005. "Adaptation: A Critique of Some Current Evolutionary Thought." *Quarterly Review of Biology* 80:19–26.

Culver, D. C. 1982. *Cave Life: Evolution and Ecology*. Cambridge, Mass.: Harvard University Press.

Culver, D. C., T. C. Kane, and D. W. Fong. 1995. *Adaptation and Natural Selection in Caves: The Evolution of Gammarus minus*.

Cummins, D. D. 1996. "Evidence for the Innateness of Deontic Reasoning." *Mind and Language* 11:160–190.

Cummins, D. D., and C. Allen, eds. 1998. *The Evolution of Mind*. Oxford: Oxford University Press.

Currie, J., and K. Sterelny. 1999. "How to Think about the Modularity of Mind-Reading." *Philosophical Quarterly* 50:145–160.

Cuvier, G. 1812. *Recherches sur les ossemens fossils des quadrupèdes, où l'on rétablit les caractèrs de plusieurs espèces d'animaux que le revolutions du globe pariossent avoir détruites*, 4 volumes. Paris: Détervville.

Daly, M., M. Wilson, and S. J. Weghorst. 1982. "Male Sexual Jealousy." *Ethology and Sociobiology* 3:11–27.

Darwin, C. 1842. "Sketch of 1842." In C. Darwin and A. R. Wallace, *Evolution by Means of Natural Selection*. Cambridge: Cambridge University Press, 1958. Reprinted in T. F. Glick and D. Kohn, eds., *Charles Darwin on Evolution*. Indianapolis: Hackett, 1996, 89–98.

Darwin, C. 1964 (1859). *On the Origin of Species*. A Facsimile of the First Edition, edited with an introduction by Ernst Mayr. Cambridge, Mass.: Harvard University Press.

Darwin, C. 1871. *The Descent of Man, and Selection in Relation to Sex*. J. T. Bonner and R. M. May, eds. Princeton: Princeton University Press, 1981.

Davies, P. S. 1996. "Discovering the Functional Mesh: On the Methods of Evolutionary Psychology." *Minds and Machines* 6:559–585.

Davies, P. S. 1999. "The Conflict of Evolutionary Psychology." In Hardcastle 1999, 67–81.

Davies, P. S., J. H. Fetzer, and T. R. Foster. 1995. "Logical Reasoning and Domain Specificity." *Biology and Philosophy* 10:1–37.

Davies, P. S. 2001. *Norms of Nature: Naturalism and the Nature of Functions.* Cambridge, Mass.: MIT Press.

Dawkins, R. 1976. *The Selfish Gene.* Oxford: Oxford University Press.

Dawkins, R. 1982. *The Extended Phenotype.* San Francisco: W. H. Freeman.

Dawkins, R. 1986. *The Blind Watchmaker.* Oxford: Oxford University Press.

Day, T., J. Pritchard, and D. Schluter. 1994. "A Comparison of Two Sticklebacks." *Evolution* 48:1723–1734.

Deacon, T. W. 1992a. "The Human Brain." In Jones, Martin, and Pilbeam 1992, 115–123.

Deacon, T. W. 1992b. "Biological Aspects of Language." In Jones, Martin, and Pilbeam 1992, 128–133.

Deacon, T. W. 1997a. *The Symbolic Species: The Co-evolution of Language and the Brain.* New York: W. W. Norton.

Deacon, T. W. 1997b. "What Makes the Human Brain Different?" *Annual Review of Anthropology* 26:337–357.

Dennett, D. C. 1983. "Intentional Systems in Cognitive Ethology: The 'Panglossian Paradigm' Defended." *Behavioral and Brain Sciences* 11:495–546.

Dennett, D. C. 1988. "Quining Qualia." In A. Marcal and E. Bisiach, eds., *Consciousness in Contemporary Science.* Oxford: Oxford University Press. Reprinted in W. Lycan, ed., *Mind and Cognition: A Reader*, S19–S47. Cambridge, Mass.: MIT Press, 1990.

Dennett, D. C. 1995. *Darwin's Dangerous Idea: Evolution and the Meanings of Life.* New York: Simon and Schuster.

Dennett, D. C. 1997. "Darwinian Fundamentalism: An Exchange." *New York Review of Books* 44:64–65.

Desmond, A. 1984. *Archetypes and Ancestors.* Chicago: University of Chicago Press.

Desmond, A. 1994. *Huxley: From Devil's Disciple to Evolution's High Priest.* New York: Addison-Wesley.

Desmond, A., and J. Moore. 1991. *Darwin: The Life of a Tormented Evolutionist.* New York: Warner Books.

Donovan, S. K., and C. R. C. Paul, eds. 1998. *The Adequacy of the Fossil Record.* Chichester, U.K.: Wiley.

Downes, S. M. 2001. "Some Recent Developments in Evolutionary Approaches to the Study of Human Cognition and Behavior." *Biology and Philosophy* 16:5575–5595.

Dupré, J., ed. 1987. *The Latest on the Best.* Cambridge, Mass.: MIT Press.

Dupré, J. 2001. *Human Nature and the Limits of Science.* Oxford: Oxford University Press.

Einhorn, H. J., and R. M. Hogarth. 1978. "Confidence in Judgment: Persistence of the Illusion of Validity." *Psychological Review* 85:395–416.

Einhorn, H. J., and R. M. Hogarth. 1981. "Behavioral Decision Theory: Processes of Judgment and Choice." *Annual Review of Psychology* 32:53–88.

Eldredge, N. 1983. "A la recherche du Docteur Pangloss." *Behavioral and Brain Sciences* 6:361–362.

Eldredge, N. 2000. *The Triumph of Evolution and the Failure of Creationism.* New York: W. H. Freeman.

Emerson, S. B., and P. A. Hastings. 1998. "Morphological Correlations in Evolution: Consequences for Phylogenetic Analysis." *Quarterly Review of Biology* 73:141–162.

Endler, J. A. 1986. *Natural Selection in the Wild.* Princeton: Princeton University Press.

Endler, J. A., and T. McLellan. 1988. "The Processes of Evolution: Towards a Newer Synthesis." *Annual Review of Ecology and Systematics* 19:395–421.

Evans, J. St. B. T. 1989. *Bias in Human Reasoning: Causes and Consequences.* London: Lawrence Erlbaum.

Evans, J. St. B. T., and D. E. Over. 1996. *Rationality and Reasoning.* Hove, England: Psychology Press.

Faith, D. 1989. "Homoplasy as Pattern." *Cladistics* 5:235–258.

Farris, J. S. 1983. "The Logical Basis of Phylogenetic Analysis." *Advances in Cladistics* 2:7–36.

Feduccia, A. 1985. "On Why the Dinosaurs Lacked Feathers." In M. K. Hecht et al., eds., *The Beginnings of Birds.* Eichstätt: Freunde des Jura Museums, 75–79.

Feduccia, A. 1993. "Aerodynamic Model for the Early Evolution of Feathers Provided by *Propithecus* (Primates, Lemuridae)." *Journal of Theoretical Biology* 160:159–164.

Feduccia, A. 1999. *The Origin and Evolution of Birds.* New Haven: Yale University Press.

Felsenstein, J. 1982. "Numerical Methods for Inferring Evolutionary Trees." *Quarterly Review of Biology* 57:397–404.

Fisher, D. 1975. "Swimming and Burrowing in *Limulus* and *Mesolimulus.*" *Fossils and Strata* 4:281–290.

Fisher, R. A. 1930. *The Genetical Theory of Natural Selection.* Oxford: Clarendon Press.

Flanagan, O. J. 1984. *The Science of the Mind.* Cambridge, Mass.: MIT Press.

Fleagle, J. C. 1999. *Primate Adaptation and Evolution,* 2nd edition. London: Academic Press.

Fodor, J. 1985. *The Modularity of Mind.* Cambridge, Mass.: MIT Press.

Fong, D. W. 1989. "Morphological Evolution of the Amphipod *Gammarus minus* in Caves: Quantitative Genetic Analysis." *American Midland Naturalist* 121:361–378.

Foster, M., and E. Ray Lancaster, eds. 1898–1902 *Scientific Memoirs of Thomas Henry Huxley.* 4 vols. London: Macmillan.

Freckleton, R. P, P. H. Harvey, and M. Pagel. 2002. "Phylogenetic Analysis and Comparative Data: A Test and Review of the Data." *American Naturalist* 160:712–726.

Frumhoff, P. C., and H. K. Reeve. 1994. "Using Phylogenies to Test Hypotheses of Adaptation: A Critique of Some Current Proposals." *Evolution* 48:172–180.

Futuyma, D. J. 1983. *Science on Trial: The Case against Creationism.* New York: Pantheon Books.

Gannon, P. J., R. W. Holloway, D. C. Broadfield, and A. R. Braun. 1998. "Asymmetry of Chimpanzee Planum Temporale: Humanlike Pattern of Wernicke's Brain Language Area Homologue." *Science* 279:220–222.

Ghiselin, M. 1983. "Lloyd Morgan's Canon in Evolutionary Context." *Behavioral and Brain Sciences* 6:362–363.

Gifford, F. 1997. "Why Not Spandrel Rather than Adaptation?" Paper presented to the American Philosophical Association.

Gigerenzer, G. 1998. "Ecological Intelligence: An Adaptation for Frequencies." In Cummins and Allen 1998, 9–29.

Gigerenzer, G., and K. Hug. 1992. "Domain Specific Reasoning: Social Contracts, Cheating, and Perspective Change." *Cognition* 43:127–177.

Gillespie, J. H. 1991. *The Causes of Molecular Evolution.* New York: Oxford University Press.

Glossip, D., and J. B. Losos. 1997. "Ecological Correlates of Number of Subdigital Lamellae in Anoles." *Herpetologica* 50: 192–199.

Godfrey-Smith, P. 2001. "Three Kinds of Adaptationism." In Orzack and Sober 2001, 335–357.

Gould, S. J. 1966. "Allometry and Size in Ontogeny and Phylogeny." *Biological Reviews* 41:587–640.

Gould, S. J. 1977a. *Ontogeny and Phylogeny*. Cambridge, Mass.: Harvard University Press.

Gould, S. J. 1977b. *Ever Since Darwin*. New York: W. W. Norton.

Gould, S. J. 1981. *The Mismeasure of Man*. New York: W. W. Norton.

Gould, S. J. 1983. "The Hardening of the Modern Synthesis." In M. Grene, ed., *Dimensions of Darwinism: Themes and Counterthemes in Twentieth-Century Evolutionary Theory*. Cambridge: Cambridge University Press, 71–93.

Gould, S. J. 1987. *An Urchin in the Storm*. New York: W. W. Norton.

Gould, S. J. 2002. *The Structure of Evolutionary Theory*. Cambridge, Mass.: Harvard University Press.

Gould, S. J., and R. C. Lewontin. 1979. "The Spandrels of San Marco and the Panglossian Paradigm: A Critique of the Adaptationist Programme." *Proceedings of the Royal Society of London* B205:581–598.

Grantham, T. 2004. "Constraints and Spandrels in Gould's *Structure of Evolutionary Theory*." *Biology and Philosophy* 19:29–43.

Grantham, T., and S. Nichols. 1999. "Evolutionary Psychology: Ultimate Explanations and Panglossian Predictions." In Hardcastle 1999, 47–66.

Griffiths, P. 1996. "The Historical Turn in the Study of Adaptation." *British Journal for the Philosophy of Science* 47:511–532.

Hamilton, W. D. 1964. "The Genetical Theory of Social Behavior." *Journal of Theoretical Biology* 7:1–52.

Hardcastle, V. G., ed. 1999. *Where Biology Meets Psychology: Philosophical Essays*. Cambridge, Mass.: MIT Press.

Harvey, P. H. 1996. "Phylogenies for Ecologists." *Journal of Animal Ecology* 65:255–263.

Harvey, P. H., and M. D. Pagel. 1991. *The Comparative Method in Evolutionary Biology*. Oxford: Oxford University Press.

Hecht, M., J. H. Ostrom, G. Viohl, and P. Wellnhofer. 1985. "The Beginnings of Birds." Freunde des Jura Museums Eichstätt, Willibaldsburg.

Heilmann, G. 1926. *The Origin of Birds*. London: Whitherby.

Henle, M. 1962. "On the Relation Between Logic and Reasoning." *Psychological Review* 69:366–378.

Hennig, W. 1966. *Phylogenetic Systematics*. Urbana: University of Illinois Press.

Herre, E. A., C. A. Machado, and S. A. West. 2001. "Selective Regime and Fig Wasp Sex Ratios: Towards Sorting Rigor from Pseudo-Rigor in Tests of Adaptation." In S. Orzack and E. Sober, eds., *Adaptation and Optimality*. Cambridge: Cambridge University Press, 191–218.

Hilton, D. J. 1995. "The Social Context of Reasoning: Conversational Inference and Rational Judgment." *Psychological Bulletin* 118:248–271.

Hodge, M. J. S. 1977. "The Structure and Strategy of Darwin's Long Argument." *The British Journal for the History of Science* 10:237–246.

Hodge, M. J. S. 1990. "Natural Selection as a Causal, Empirical, and Probabilistic Theory." In L. Krüger, G. Gigerenzer, and M. S. Morgan, eds., *The Probabilistic Revolution*. Vol. 2, *Ideas in the Sciences*. Cambridge, Mass.: MIT Press, 233–270.

Holden, C. 2004. "The Origin of Speech." *Science* 303:1316–1319.

Holling, P. S. 1964. "The Analysis of Complex Population Processes." *Canadian Entomologist* 96:345–347.

Holloway, R. 1979. "Brain Size, Allometry, and Reorganization: Toward a Synthesis." In M. Hahn, C. Jensen, and B. Dudek, eds., *Development and Evolution of Brain Size*, 59–88. New York: Academic Press.

Holloway, R. 1983. "Human Paleontological Evidence Relevant to Language Behavior." *Human Neurobiology* 2:105–114.

Hull, D. L. 1988. *Science as a Process*. Chicago: University of Chicago Press.

Hull, D. L. 1978. "Planck's Principle." *Science* 202:717–723.

Hume, D. 1777. *Enquiries Concerning Human Understanding and Concerning the Principles of Morals*. Ed. L. A. Selby-Bigge and rev. by P. H. Nidditch. Oxford: Clarendon Press, 1975.

Huxley, T. H. 1863. *Man's Place in Nature*. London: Williams and Norgat.

Huxley, T. H. 1868. "On the Animals." In Foster and Lancaster 1898–1902, 3:303–366.

Huxley, T. H. 1871. "Administrative Nihilism." In T. H. Huxley, *Collected Essays,* vol. 7. New York: Greenwood, 251–289.

Huxley, T. H. 1888. "The Struggle for Existence in Human Society." In T. H. Huxley, *Collected Essays*, volume 9. New York: D. Appleton, 195–236..

Huxley, T. H. 1893. "Evolution and Ethics." In T. H. Huxley, *Collected Essays*, vol. 9. New York: Greenwood, 46–116.

Huxley, T. H. 1894. "Evolution and Ethics (prolegomena)." In T. H. Huxley, *Collected Essays*, vol. 9. New York: Greenwood, 1–45.

Irschick, D. J., C. C. Austin, K. Petren, R. N. Fisher, J. B. Losos, and O. Ellers. 1996. "A Comparative Analysis of Clinging Ability among Pad-Bearing Lizards." *Biological Journal of the Linnean Society* 59:21–35.

Jackman, T. R., A. Larson, K. de Queiroz, and J. B. Losos. 1999. "Phylogenetic Relationships and the Tempo of Early Diversification in Anolis lizards." *Systematic Biology* 48:254–285.

Jain, S. K., and A. D. Bradshaw. 1966. "Evolutionary Divergence among Adjacent Plant Populations. I. The Evidence and Its Theoretical Analysis." *Heredity* 21:407–441.

Jerison, H. 1973. *The Evolution of Intelligence*. New York: Academic Press.

Jones, R., D. C. Culver, and T. C. Kane. 1992. "Are Parallel Morphologies of Cave Organisms the Result of Similar Selection Pressures?" *Evolution* 46:353–365.

Jones, S., R. Martin, and D. Pilbeam. 1992. *The Cambridge Encyclopedia of Human Evolution*. Cambridge: Cambridge University Press.

Kamil, A. C., J. R. Krebs, and H. R. Pulliam, eds. 1987. *Foraging Behavior*. New York: Plenum Press.

Kane, T. C., and R. C. Richardson. 2005. "Natural Selection." In D. C. Culver and W. B. White, eds., *Encyclopedia of Caves*. Amsterdam: Elsevier, 409–411.

Kauffman, S. A. 1993. *The Origins of Order*. Oxford: Oxford University Press.

Ketelaar, T., and B. J. Ellis. 2000. "Are Evolutionary Explanations Unfalsifiable? Evolutionary Psychology and the Lakatosian Philosophy of Science." *Psychological Inquiry* 11:1–21.

Kingsland, S. 1985. *Modeling Nature*. Chicago: University of Chicago Press.

Kingsolver, J. G., H. E. Hoekstra, J. M. Hoekstra, D. Berrigan, S. N. Vignieri, C. E. Hill, A. Hoang, P. Gilbert, and P. Beerl. 2001. "The Strength of Phenotypic Selection in Natural Populations." *American Naturalist* 157:245–261.

Kitcher, P. 1982. *Abusing Science: The Case Against Creationism*. Cambridge, Mass.: MIT Press.

Kitcher, P. 1985. *Vaulting Ambition: Sociobiology and the Quest for Human Nature*. Cambridge, Mass.: MIT Press.

Kitcher, P. 2003. "Darwin's Achievement." In Kitcher, *In Mendel's Mirror.* Oxford: Oxford University Press, 45–93.

Kitcher, P., and A. L. Vickers. 2002. "Pop Sociobiology Reborn: The Evolutionary Psychology of Sex and Violence." In Cheryl Travis, ed., *Evolution, Gender, and Rape*. Cambridge, Mass.: MIT Press, 139–168.

Kluge, A. G. 1983. "Cladistics and the Classification of the Great Apes." In R. L. Ciochon and R. S. Corruccini, eds., *New Interpretations of Ape and Human Ancestry*. New York: Plenum Publishing, 151–177.

Knight, F. 1921. *Risk, Uncertainty, and Profit.* Boston: Houghton Mifflin.

Kitcher, P. 1987. "Why Not the Best?" In John Dupré 1987, 77–102.

Kohn, D., ed. 1985. *The Darwinian Heritage.* Princeton: Princeton University Press.

Kottler, J. J. 1985. "Charles Darwin and Alfred Russel Wallace: Two Decades of Debate over Natural Selection." In Kohn 1985, 367–432.

Krebs, J. R., J. T. Erichsen, M. I. Webber, and E. L. Charnov. 1977. "Optimal Prey Selection in the Great Tit (*Parus major*)." *Animal Behavior* 25:30–38.

Krebs, J. R., and N. Davies. 1978. *Behavioural Ecology: An Evolutionary Approach.* Oxford: Blackwell.

Krebs, J. R., and N. Davies. 1981. *An Introduction to Behavioural Ecology.* Oxford: Blackwell.

Kreitman, M. 1983. "Nucleotide Polymorphism at the Alcohol Dehydrogenase Locus of *Drosophila melanogaster*." *Nature* 304:412–417.

Kreitman, M., and R. R. Hudson. 1991. "Inferring the Histories of the *Adh* and *Adh-dup* loci in *Drosophila melanogaster* from Patterns of Polymorphism and Divergence." *Genetics* 127:565–582.

Kuhn, T. S. 1957. *The Copernican Revolution: Planetary Astronomy in the Development of Western Thought.* Cambridge, Mass.: Harvard University Press.

Lamarck, J. 1809. *Philosophie zoologique.* 2 volumes. Paris: Dentu.

Lande, R. 1976. "Natural Selection and Random Genetic Drift in Phenotypic Evolution." *Evolution* 30:314–334.

Lande, R. 1977. "Statistical Tests for Natural Selection on Quantitative Characters." *Evolution* 37:442–444.

Lande, R., and S. J. Arnold. 1983. "The Measurement of Selection on Correlated Characters." *Evolution* 37:1210–1226.

Landreth, T., and R. C. Richardson. 2004. "Localization and the New Phrenology: A Review Essay on William Uttal's *The New Phrenology*." *Philosophical Psychology* 17:107–123.

Langerhaus, R. B., and T. J. DeWitt. 2004. "Shared and Unique Features of Evolutionary Diversification." *American Naturalist* 164:335–339.

Larson, A., and J. B. Losos. 1996. "Phylogenetic Systematics of Adaptation." In Rose and Lauder 1996, 187–220.

Lauder, G. V. 1981. "Form and Function: Structural Analysis in Evolutionary Morphology." *Paleobiology* 7:430–442.

Lauder, G. 1982. "Historical Biology and the Problem of Design." *Journal of Theoretical Biology* 97:57–67.

Lauder, G. V. 1996. "The Argument from Design." In M. R. Rose and G. V. Lauder, eds., *Adaptation.* San Diego: Academic Press, 55–91.

Leroi, A. M., M. R. Rose, and G. V. Lauder. 1994. "What Does the Comparative Method Reveal about Adaptation?" *American Naturalist* 143:381–402.

Levi, I. 1983. "Who Commits the Base Rate Fallacy?" *Behavioral and Brain Sciences* 6:502–506.

Levi, I. 1996. "Fallacy and Controversy About Base Rates." *Behavioral and Brain Sciences* 19:31–32.

Levins, R. C. 1963. "Theory of Fitness in a Heterogeneous Environment. II. Developmental Flexibility and Niche Selection." *American Naturalist* 97:75–90.

Levins, R. C. 1968. *Evolution in a Changing Environment.* Princeton: Princeton University Press.

Lewens, T. 2002. "Adaptationism and Engineering." *Biology and Philosophy* 17:1–31.

Lewontin, R. C. 1957. "The Adaptation of Populations to Varying Environments." *Cold Spring Harbor Symposia on Quantitative Biology* 22:395–408.

Lewontin, R. C. 1969. "The Bases of Conflict in Biological Explanation." *Journal of the History of Biology* 2:35–45.

Lewontin, R. C. 1970. "Units of Selection." *Annual Review of Ecology and Systematics* 1:1–13.

Lewontin, R. C. 1974a. "The Analysis of Variance and the Analysis of Causes." *American Journal of Human Genetics* 26:400–411.

Lewontin, R. C. 1974b. *The Genetic Basis of Evolutionary Change.* New York: Columbia University Press.

Lewontin, R. C. 1976. "Sociobiology—a Caricature of Darwinism." *Proceedings of the Philosophy of Science Association,* volume 2. East Lansing, Mich.: Philosophy of Science Association, 22–31.

Lewontin, R. C. 1978. "Adaptation." *Scientific American* 239:212–230.

Lewontin, R. C. 1979. "Sociobiology as an Adaptationist Program." *Behavioral Science* 24:5–14.

Lewontin, R. C. 1982. *Human Diversity.* New York and San Francisco: W. H. Freeman.

Lewontin, R. C. 1983. Review of Lumsden and Wilson, *Genes, Mind, and Culture. The Sciences* (Proceedings of the New York Academy of Science).

Lewontin, R. C. 1987. "The Shape of Optimality." In Dupré 1987, 151–160.

Lewontin, R. C. 1990. "The Evolution of Cognition." In D. N. Osherson and E. E. Smith, eds., *An Invitation to Cognitive Science: Thinking.* Cambridge, Mass.: MIT Press, 229–246.

Lewontin, R. C., and S. J. Gould. 1979. "The Spandrels of San Marco and the Panglossian Paradigm." *Proceedings of the Royal Society of London* B205:581–598.

Lewontin, R. C., S. Rose, and L. Kamin. 1984. *Not in Our Genes.* New York: Pantheon.

Lieberman, P. 1991. *The Biology and Evolution of Human Language.* Cambridge, Mass.: Harvard University Press.

Lieberman, P. 1992. "Human Speech and Language." In Jones, Martin, and Pilbeam 1992, 134–137.

Lieberman, D. 1995. "Testing Hypotheses about Recent Human Evolution from Skulls: Integrating Morphology, Function, Development and Phylogeny." *Current Anthropology* 36:159–197.

Lieberman, D. E., B. A. Wood, and D. R. Pilbeam. 1996. "Homoplasy and Early *Homo*: An Analysis of the Evolutionary Relationships of *H. habilis sensu stricto* and *H. rudolfensis.*" *Journal of Human Evolution* 30:97–120.

Looren de Jong, H., and W. J. van der Steen. 1998. "Biological Thinking in Evolutionary Psychology: Rockbottom or Quicksand?" *Philosophical Psychology* 11:183–206.

Lopes, L. L. 1991. "The Rhetoric of Irrationality." *Theory and Psychology* 1:65–82.

Lorenz, K. 1966. *On Aggression.* New York: Harcourt, Brace, Jovanovich.

Losos, J. B. 1990. "The Evolution of Form and Function: Morphology and Locomotor Performance in West Indian *Anolis* Lizards." *Evolution* 44:1189–1203.

Losos, J. B. 1992. "The Evolution of Convergent Structure in Caribbean *Anolis* Communities." *Systematic Biology* 41:403–420.

Losos, J. G., T. R. Jackman, A. Larson, K. de Queiroz, and L. Rodriguez-Schettino. 1998. "Contingency and Determinism in Replicated Adaptive Radiations of Island Lizards." *Science* 279:2115–2118.

Losos, J. B., and D. B. Miles. 2002. "Testing the Hypothesis That a Clade Has Adaptively Radiated: Iguanid Lizard Clades as a Case Study." *American Naturalist* 160:147–167.

Lumsden, C., and E. O. Wilson. 1981. *Genes, Minds, and Culture.* Cambridge, Mass.: Harvard University Press.

Lyell, C. 1863. *The Geological Evidences of the Antiquity of Man, with Remarks on the Theories of the Origin of Species by Variation,* 2nd ed. London: Charles Murray.

Lynch, M. 1991. "Methods for the Analysis of Comparative Data in Evolutionary Biology." *Evolution* 45:1065–1080.

Lyons, S. L. 1999. *Thomas Henry Huxley: The Evolution of a Scientist.* Amherst, N.Y.: Prometheus Books.

MacArthur, R. H. 1957. "On the Relative Abundance of Bird Species." *Proceedings of the National Academy of Sciences* 43:293–295.

MacArthur, R. H. 1960. "On the Relative Abundance of Species." *American Naturalist* 94:25–36.

MacNair, M. 1979. "The Genetics of Copper Tolerance in the Yellow Monkey Flower, *Mimulus guttatus*. I. Crosses to Nontolerants." *Genetics* 91:553–563.

Maddison, W. P., M. J. Donaghue, and D. R. Maddison. 1984. "Outgroup Analysis and Parsimony." *Systematic Zoology* 33:83–103.

Margolis, H. 1987. *Patterns, Thinking, and Cognition*. Chicago: University of Chicago Press.

Maynard Smith, J. 1978. "Optimization Theory in Evolution." *Annual Review of Ecology and Systematics* 9:31–56.

Maynard Smith, J. 1982. *Evolution and the Theory of Games*. Cambridge: Cambridge University Press.

Maynard Smith, J. 1987. "How to Model Evolution." In J. Dupré 1987, 119–131.

Maynard Smith, J. 1995. "Genes, Memes, and Minds." *New York Review of Books*, 42.

Maynard Smith, J., R. Burian, S. Kauffman, P. Alberch, J. Campbell, B. Goodwin, R. Lande, D. Raup, and L. Wolpert. 1985. "Developmental Constraints and Evolution." *Quarterly Review of Biology* 60:265–287.

Mayr, E. 1963. *Animal Species and Evolution*. Cambridge, Mass.: Harvard University Press.

Mayr, E. 1983. "How to Carry Out the Adaptationist Program." *American Naturalist* 121:324–334.

Mayr, E. 2001. *What Evolution Is*. New York: Basic Books.

McCollum, M. A. 1999. "The Robust Australopithecine Face: A Morphogenetic Perspective." *Science* 284:301–305.

McGuffin, P., B. Riley, and R. Plomin. 2001. "Toward Behavioral Genomics." *Science* 291:1232–1249.

Mendel, G. 1865. "Experiments in Plant Hybridization." *Verhandlungen des naturforschenden Vereins* 4:3–47 (1866). Translated by C. T. Druery and W. Bateson. Available at www.mendelweb.org.

Michod, R. 1982. "The Theory of Kin Selection." *Annual Review of Ecology and Systematics* 13:23–55.

Milkman, R., ed. 1982. *Perspectives on Evolution*. Sunderland, Mass.: Sinauer.

Mitchell, S. D. 2003. *Biological Complexity and Integrative Pluralism*. Cambridge: Cambridge University Press.

Mitchell, R. W., W. H. Russell, and W. R. Elliott. 1977. "Mexican Eyeless Characin Fishes, Genus *Astyanx*: Environment, Distribution, and Evolution." Special Publication No. 11. Lubbock: Texas Tech University.

Morley, J. 1871. "Mr. Darwin on the Descent of Man." *Times* (London). April 7.

Morris, D. 1967. *The Naked Ape*. New York: McGraw-Hill.

Nesse, R. M. 1990. "Evolutionary Explanations of Emotions." *Human Nature* 1:261–289.

Nozick, R. 1993. *The Nature of Rationality*. Princeton: Princeton University Press.

Oaksford, M., and N. Chater. 1992. "Bounded Rationality in Taking Risks and Drawing Inferences." *Theory and Psychology* 2:225–230.

Oaksford, M., and N. Chater. 1993. "Reasoning Theories and Bounded Rationality." In Manktelow and D. Over, eds., *Rationality: Psychological and Philosophical Perspectives*. London: Routledge, 31–60.

Oaksford, M., and N. Chater. 1994. "A Rational Analysis of the Selection Task as Optimal Data Selection." *Psychological Review* 101:608–631.

Oaksford, M., and N. Chater. 1995. "Theories of Reasoning and the Computational Explanation of Everyday Inference." *Thinking and Reasoning* 1:121–152.

Oaksford, M., and N. Chater. 1996. "Rational Explanation of the Selection Task." *Psychological Review* 103:381–391.

Orgel, L. E. 2004. "Prebiotic Chemisty and the Origin of the RNA World." *Critical Reviews in Biochemistry and Molecular Biology* 39:99–123.

Orzack, S., and E. Sober. 1994a. "Optimality Models and the Test of Adaptationism." *American Naturalist* 143:361–380.

Orzack, S., and E. Sober. 1994b. "How (not) to Test an Optimality Model." *Trends in Ecology and Evolution* 9:265–267.

Orzack, S., and E. Sober. 1996. "How to Formulate and Test Adaptationism." *American Naturalist* 148:202–210.

Orzack, S., and E. Sober, eds. 2001. *Adaptation and Optimality*. Cambridge: Cambridge University Press.

Ospovat, D. 1981. *The Development of Darwin's Theory*. Cambridge: Cambridge University Press.

Ostrom, J. H. 1974. "*Archaeopteryx* and the Origin of Flight." *Quarterly Review of Biology* 49:27–47.

Ostrom, J. H. 1976. *Archaeopteryx* and the Origin of Birds." *Biological Journal of the Linnean Society* 8:91–182.

Owen, R. 1862. "On the Archaeopteryx of Von Meger with a Description of the Fossil Remains of a Longtailed Bird from the Lithographic Slate of Solenhofen." *Philosophical Transactions of the Royal Society* 153:33–47.

Padian, K. 2001. "Cross-Testing Adaptive Hypotheses: Phylogenetic Analysis and the Origin of Bird Flight." *American Zoologist* 41:598–607.

Page, R. E., Jr., and S. D. Mitchell. 1991. "Self Organization and Adaptation in Insect Societies." In A. Fine, M. Forbes, and L. Wessels, eds., *PSA 1989*, vol. 2. East Lansing, Mich.: Philosophy of Science Association, 289–298.

Parker, G. A., and J. Maynard Smith. 1990. "Optimality Theory in Evolutionary Biology." *Nature* 348:27–33.

Parkes, K. C. 1966. "Speculations on the Origin of Feathers." *Living Bird* 5:77–86.

Piatelli-Palmarini, M. 1994. *Inevitable Illusions: How Mistakes of Reason Rule Our Minds*. New York: Wiley.

Pinker, S. 1994. *The Language Instinct: How the Mind Creates Language*. New York: William Morrow.

Pinker, S. 1997. *How the Mind Works*. Cambridge, Mass.: MIT Press.

Pinker, S., and P. Bloom. 1992. "Natural Language and Natural Selection." *Behavioral and Brain Sciences* 13(1990):707–784. Reprinted in Barkow, Cosmides, and Tooby 1992, 451–493.

Popper, K. 1959. *The Logic of Scientific Discovery*. New York: Basic Books.

Poulson, T. L. 1963. "Cave Adaptation in Amblyopsid Fishes." *American Midland Naturalist* 70:257–290.

Poulson, T. L., and W. B. White. 1969. "The Cave Environment." *Science* 165:971–981.

Profet, M. 1992. "Pregnancy Sickness as an Adaptation: A Deterrent to Maternal Ingestion of Teratogens." In Barkow, Cosmides, and Tooby 1992, 327–365.

Prum, R. O., and A. H. Brush. 2002. "The Evolutionary Origin and Diversification of Feathers." *Quarterly Review of Biology* 77:261–295.

Pyke, G. H. 1984. "Optimal Foraging Theory: A Critical Review." *Annual Review of Ecology and Systematics* 15:523–575.

Pyke, G. H., H. R. Pullian, and E. L. Charnov. 1977. "Optimal Foraging: A Selective Review of Theory and Tests." *Quarterly Review of Biology* 52:137–154.

Raff, R. A. 1996. *The Shape of Life*. Chicago: University of Chicago Press.

Reeve, H. K., and P. W. Sherman. 1993. "Adaptation and the Goals of Evolutionary Research." *Quarterly Review of Biology* 68:1–32.

Richards, R. J. 1987. *Darwin and the Emergence of Evolutionary Theories of Mind and Behavior*. Chicago: University of Chicago Press.

Richardson, R. C. 1996. "Critical Review of Robert Brandon, *Adaptation and Environment*." *Philosophy of Science* 63:122–136.

Richardson, R. C. 2000. "The Prospects for an Evolutionary Psychology: Human Language and Human Reasoning." *Minds and Machines* 6:541–557.

Richardson, R. C. 2001a. "Evolution without History: Critical Reflections on Evolutionary Psychology." In H. R. Holcomb, ed., *Conceptual Challenges in Evolutionary Psychology: Innovative Research Strategies*. Kluwer Academic Publishers, 327–373.

Richardson, R. C. 2001b. "Complexity, Self Organization, and Selection." *Biology and Philosophy* 16:655–683.

Richardson, R. C. 2003a. "Epicycles and Explanations in Evolutionary Psychology." *Psychological Inquiry* 11:46–49.

Richardson, R. C. 2003b. "Engineering Design and Adaptation." *Philosophy of Science* 70:1–12.

Richardson, R. C. 2003c. "Adaptation, Adaptationism, and Optimality." *Biology and Philosophy* 18:695–713.

Richardson, R. C., and R. M. Burian. 1992. "A Defense of Propensity Interpretations of Fitness." In M. Forbes and D. Hull, eds., *PSA 1992*, vol. 1. East Lansing, Mich.: Philosophy of Science Association, 349–362.

Ricklefs, R. E. 1987. "Community Diversity: Relative Roles of Local and Regional Processes." *Science* 235:167–171.

Ridley, M. 1993. *The Red Queen: Sex and the Evolution of Human Nature*. New York: Perennial Press.

Rose, H. R., and S. Rose, eds. 2000. *Alas, Poor Darwin: Arguments Against Evolutionary Psychology*. New York: Harmony Books.

Rose, M. R., and G. V. Lauder, eds. 1996. *Adaptation*. New York: Academic Press.

Rosenberg, A. 2001. "Reductionism in a Historical Science." *Philosophy of Science* 68:135–163.

Ruben, J. 1991. "Reptilian Physiology and the Flight Capacity of *Archaeopteryx*." *Evolution* 45:1–17.

Ruse, M. 1979a. *Sociobiology: Sense or Nonsense?* Dordrecht: D. Reidel.

Ruse, M. 1979b. *The Darwinian Revolution: Science Red in Tooth and Claw*. Chicago: University of Chicago Press.

Russell, D. A. 1972. "Ostrich Dinosaurs from the Late Cretaceous of Western Canada." *Canadian Journal of Earth Sciences* 9:375–402.

Sahlins, M. 1976. *The Use and Abuse of Biology*. Ann Arbor: University of Michigan Press.

Schank, J. C., and W. C. Wimsatt, 1988. "Generative Entrenchment and Evolution." In A. Fine and P. K. Machamer, eds., *PSA 1986*. Vol. 2. East Lansing, Mich.: The Philosophy of Science Association, 33–60.

Schluter, D., E. A. Clifford, M. Nemethy, and J. S. McKinnon. 2004. "Parallel Evolution and Inheritance of Quantitative Traits." *American Naturalist* 163:S809–S822.

Schoener, T. W. 1987. "A Brief History of Optimal Foraging Ecology." In Kamil, Krebs, and Pulliam 1987, 6–67.

Shapiro, L. A. 1999. "Presence of Mind." In Hardcastle 1999, 83–98.

Shapiro, R. 2006. "Small Molecule Interactions Were Central to the Origin of Life." *Quarterly Review of Biology* 81:105–125.

Shepard, R. N. 1992. "The Perceptual Organization of Colors: An Adaptation to Regularities of the Terrestrial World?" In Barkow, Cosmides, and Tooby 1992, 495–532.

Sherman, P. W. 1989. "The Clitoris Debate and the Levels of Analysis." *Animal Behaviour* 37:697–698.

Sherman, P. W. 1998. "The Levels of Analysis" *Animal Behaviour* 36:616–619.

Sibley, C. G., and J. E. Ahlquist. 1984. "The Phylogeny of the Hominid Primates as Indicated by DNA-DNA Hybridization." *Journal of Molecular Evolution* 20:2–15.

Skelton, R. R., and H. M. McHenry. 1992. "Evolutionary Relationships among Early Hominids." *Journal of Human Evolution* 23:309–349.

Skelton, R. R., and H. M. McHenry. 1997. "Trait List Bias and a Reappraisal of Early Hominid Phylogeny." *Journal of Human Evolution* 34:109–113.

Sloman, S. A. 1996. "The Empirical Case for Two Systems of Reasoning." *Psychological Bulletin* 119:3–22.

Smith, A. 1759. *The Theory of the Moral Sentiments*. London: Miller; Edinburgh: Kincaid and Bell.

Sneath, P. H. A., and R. R. Sokal. 1973. *Numerical Taxonomy: The Principles and Practice of Numerical Classification*. San Francisco: W. H. Freeman.

Sober, E. 1984. *The Nature of Selection: Evolutionary Theory in Philosophical Focus*. Cambridge, Mass.: MIT Press.

Sober, E. 1987. "What Is Adaptationism?" In Dupré 1987, 105–118.

Spencer, H. 1851. *Social Statics: Or, the Conditions Essential to Human Happiness Specified, and the First of Them Developed*. London: Chapman.

Spencer, H. 1852. "The Development Hypothesis." *The Leader.* Reprinted in Spencer 1901, vol. 1, 1–7.

Spencer, H. 1893/1978. *The Principles of Ethics*. Indianapolis: Liberty Classics.

Spencer, H. 1901. *Essays: Scientific, Political, and Speculative*. Williams and Norgate.

Stein, E. 1996. *Without Good Reason: The Rationality Debate in Philosophy and Cognitive Science*. Oxford: Oxford University Press.

Steiner, H. 1917. "Das Problem der Diastataxie des Vogelflügels." *Jenaische Zeitschrift für Naturwissenschaft* 55:222–496.

Sterelny, K., and P. E. Griffiths. 1999. *Sex and Death: An Introduction to Philosophy of Biology*. Chicago: Chicago University Press.

Stich, S. P. 1993. "Naturalizing Epistemology: Quine, Simon, and the Prospects for Pragmatism." In C. Hookway and D. Peterson, eds., *Philosophy and Cognitive Science*. Cambridge: Cambridge University Press, 1–17.

Strait, D. S., F. E. Grine, and M. A. Moniz. 1997. "A Reappraisal of Early Hominid Phylogeny." *Journal of Human Evolution* 32:17–82.

Sultan, S. E., and F. A. Bazzaz. 1993a. "Phenotypic Plasticity in *Polygonum persicaria*. I. Diversity and Uniformity in Genotypic Norms of Reaction to Light." *Evolution* 47:1009–1031.

Sultan, S. E., and F. A. Bazzaz. 1993b. "Phenotypic Plasticity in *Polygonum persicaria*. II. Norms of Reaction to Soil Moisture and the Maintenance of Genetic Diversity." *Evolution* 47:1032–1049.

Sultan, S. E., and F. A. Bazzaz. 1993c. "Phenotypic Plasticity in *Polygonum persicaria*. III. The Evolution of Ecological Breadth for Nutrient Requirement." *Evolution* 47:1050–1071.

Symons, D. 1979. *The Evolution of Human Sexuality*. Cambridge, Mass.: Harvard University Press.

Symons, D. 1992. "On the Use and Misuse of Darwinism in the Study of Human Behavior." In Barkow, Cosmides, and Tooby 1992, 137–159.

Templeton, A. 1982. "Adaptation and the Integration of Evolutionary Forces." In Milkman 1982, 15–31.

Thagard, P. 1982. "From the Descriptive to the Normative in Philosophy and Logic." *Philosophy of Science* 49:250–267.

Thagard, P., and R. E. Nisbett. 1983. "Rationality and Charity." *Philosophy of Science* 50:250–267.

Thaler, R. H. 1980. "Toward a Positive Theory of Consumer Choice." *Journal of Economic Behavior and Organization* 1:39–60.

Thaler, R. H. 1992. *The Winner's Curse: Paradoxes and Anomalies of Economic Life.* New York: The Free Press.

Thornhill, R. 1990. "The Study of Adaptation." In M. Bekoff and D. Jamieson, eds., *Interpretation and Explanation in the Study of Behavior.* vol. 2. Westview Press, 31–62.

Thornhill, R., and C. Palmer. 2000. *A Natural History of Rape: Biological Bases of Sexual Coercion.* Cambridge, Mass.: MIT Press.

Tooby, J., and L. Cosmides. 1992. "The Psychological Foundations of Culture." In Barkow, Cosmides, and Tooby 1992, 19–136.

Tooby, J., and I. De Vore. 1987. "The Reconstruction of Hominid Evolution Through Strategic Modeling." In W. G. Kinzey, ed., *The Evolution of Human Behavior: Primate Models.* Albany: SUNY Press, 183–237.

Trivers, R. L. 1971. "The Evolution of Reciprocal Altruism." *Quarterly Review of Biology* 46:35–57.

Trivers, R. L. 1972. "Parental Investment and Sexual Selection." In B. Campbell, ed., *Sexual Selection and the Descent of Man.* Chicago: Aldine, 136–179.

Uttal, W. R. 2001. *The New Phrenology: The Limits of Localizing Cognitive Processes in the Brain.* Cambridge, Mass.: MIT Press.

van den Berghe, P. 1980. "Incest and Exogamy: A Sociobiological Reconsideration." *Behavioral and Brain Sciences* 5:27.

van den Berghe, P. 1983. "Human Inbreeding Avoidance: Culture in Nature." *Behavioral and Brain Sciences* 6:91–123.

Vogel, S. 1988. *Life's Devices: The Physical World of Animals and Plants.* Princeton: Princeton University Press.

Vogel, S. 1998. *Cat's Paws and Catapults: Mechanical Worlds of Nature and People.* New York: W. W. Norton.

Wallace, A. R. 1864. "The Origin of Human Races and the Antiquity of Man Deduced from the Theory of 'Natural Selection.'" *Anthropological Review* 2:158–187.

Ward, S. 1999. "*Equatorius*: A New Homonoid Genus from the Middle Miocene of Kenya." *Science* 285:1382–1386.

Wetherick, N. E. 1993. "Human Rationality." In K. I. Manktelow and D. Over, eds., *Rationality: Psychological and Philosophical Perspectives.* London: Routledge, 83–109.

Wetherick, N. E. 1995. "Reasoning and Rationality: A Critique of Some Experimental Paradigms." *Theory and Psychology* 5:429–448.

White, T. D., et al. 2003. "Pleistocene *Homo sapiens* from Middle Awash, Ethiopia." *Nature* 423:742–747.

Wilkins, H. 2005. "Fish." In D. C. Culver and W. B. White, eds., *Encyclopedia of Caves.* Amsterdam: Elsevier, 241–250.

Williams, G. C. 1966. *Adaptation and Natural Selection.* Princeton: Princeton University Press.

Williams, G. C. 1992. *Natural Selection: Domains, Levels, and Challenges.* Princeton: Princeton University Press.

Wilson, E. O. 1975. *Sociobiology: The New Synthesis.* Cambridge, Mass.: Harvard University Press.

Wilson, E. O. 1976. "Academic Vigilantism and the Political Significance of Sociobiology." *BioScience* 26:183. In Caplan 1978, 291–303.

Wilson, E. O. 1978. *On Human Nature.* Cambridge, Mass.: Harvard University Press.

Wimsatt, W. C. 2007. *Re-engineering Philosophy for Limited Beings: Piecewise Approximations to Reality.* Cambridge, Mass.: Harvard University Press.

Wimsatt, W. C., and J. C. Schank. 1988. "Two Constraints on the Evolution of Complex Adaptations and the Means for Their Avoidance." In M. Nitecki, ed., *Evolutionary Progress.* Chicago: University of Chicago Press, 231–273.

Wood, B., and M. Collard. 1999. "The Human Genus." *Science* 284:65–71.

Xu, X., A. Tang, and X. Wang. 1999a. "A Therizinosauroid Dinosaur with Integumentary Structures from China." *Nature* 401:262–266.

Xu, X., A. Tang, and X. Wang. 1999b. "A Dromaeosaurid Dinosaur with a Filamentous Integument from the Yixian Formation of China." *Nature* 401:200–204.

Xu, X., Z. Zhou, and R. O. Prum. 2001. "Branched Integumental Structures in *Sinornithosaurus* and the Origin of Feathers." *Nature* 410:200–204.

Index